PALEOLITHIC SITE OF THE DOUARA CAVE AND PALEOGEOGRAPHY OF PALMYRA BASIN IN SYRIA

PALEOLITHIC SITE OF THE DOUARA CAVE AND PALEOGEOGRAPHY OF PALMYRA BASIN IN SYRIA

Part I: Stratigraphy and Paleogeography in the Late Quaternary

Edited by
Kazuro HANIHARA and Yutaka SAKAGUCHI

UNIVERSITY OF TOKYO PRESS

The University Museum, The University of Tokyo, Bulletin No. 14, 1978

UTP 3044-68824-5149

Printed in Japan

ISBN 0-86008-209-1

CONTENTS

CHAPTER 1
Introduction

Kazuro Hanihara

Department of Anthropology, Faculty of Science, The University of Tokyo

The Douara Cave is located about 200 km northeast of Damascus, Syria, and about 18 km northeast of Palmyra which is known as a famous oasis town. The cave is on the southern slope of Jabal ad Douara in the mountain masses that bound the Palmyra Basin on the north. These mountains are part of the southern Palmyra Mountains which branch in a northeasterly direction from the southeast of the Anti-Lebanon Mountains. The floor of the cave is about 557 m above sea level. The entrance is 10 m wide and 12 m high, and the depth is 14 m. The cave has a corridor-like plan.

The Douara Cave was discovered by the Third Tokyo University Scientific Expedition to Western Asia, 1967, during a general prehistoric survey of Syria and Lebanon. Later, in 1970, members of the Fourth Expedition carried out a systematic excavation of the cave for 2 months from August through September, and confirmed that it was rich in stone artifacts and fossil animal bones derived from the Middle and Epi-Paleolithic ages. In this season, however, time allowed us to excavate only a minor part of the cave. The results of the excavation in this season have been reported under the editorship of Suzuki and Takai (1973, 1974), in which the background of the expedition and the background of the survey of the Douara Cave were described in detail.

In order to make a complete excavation of the cave, the Fifth Tokyo University Scientific Expedition to Western Asia was organized in 1974 with the following members:

Kazuro Hanihara	Director of the Expedition, Physical Anthropology
Banri Endo	Physical Anthropology
Yutaka Sakaguchi	Physical Geography
Kunihiko Endo	Physical Geography
Takeei Koizumi	Physical Geography
Tsuyoshi Fujimoto	Prehistory
Takeru Akazawa	Prehistory

The main purposes of this field season were to obtain artificial as well as natural remains from the Paleolithic deposits which covered the whole of the cave floor, and to reconstruct the natural environments and paleogeography around the cave.

The excavation and geographical surveys were carried out for 4 months during the period from August through November, 1974, and, very fortunately, a tremendous number of prehistoric remains were obtained. At the same time, paleogeographic data accumulated in this season made possible a reasonable reconstruction of the natural environments in prehistoric ages.

The field work concentrated mostly on the excavation of the cave deposits, as in the preceding season, in order to expose the complete sequence of cave deposits from the surface to the bed rock, and to examine the industrial succession according to the depositional sequence of the cave.

Through partial excavations of the cave in the 1970 season, it was found that the cave had a deposit more than 4 m in depth, divided into several geologically and culturally defined layers with two distinct lithic assemblages: Middle Paleolithic from the lower horizon and Epi-Paleolithic from the upper horizon. These findings have assured the particular importance of this cave in studies of the evolution of the Paleolithic age at a single site. But the 1974 season's data brought to light yet another important fact. The sequence of Middle Paleolithic assemblages in the 1.5 m of the lower horizon manifests significant differences, both typologically and technologically. Those new data are on the way to becoming some of the most important evidence for examining the meaning of the variability seen in the Middle Paleolithic assemblages of the Levant.

In association with the excavation work, geological surveys in the vicinity of the cave were also undertaken. During the surveys, mainly flint-strewn fields, characterized by the presence of artifactual flints, were found in the Palmyra and Douara Basins. Thousands of pieces classifiable as artifacts were collected from these localities. Through examination of these samples, it was found that the flint artifacts fall definitely into two main groups: one Paleolithic and the other Pre-Pottery and Pottery Neolithic. These new findings may provide new data for making a framework to which chronological work on the prehistory of the Syrian desert can be related.

In addition, one result which will be emphasized is that several flint factory sites were recovered in the Douara Basin, which is surrounded by the Douara Mountains and located on the opposite side of the Douara Cave. The sites yielded various types of industries which belong to the Lower Paleolithic, Mounsterian and Pre-Pottery Neolithic ages. The distance between the Douara Cave and Locality 38, one of the factory sites which yielded the same type of flint assemblage, is only about 1 km in a straight line, across a ridge. Moreover, around the slopes of the Douara Mountain and in many caves and shelters in the area, a scattering of Pre-Pottery Neolithic type flints were found. Evidence that two distinct kinds of sites, one the cave sites and the other the factory sites, were located in a restricted area during the same age may provide basic information for the reconstruction of behavioral patterns of prehistoric peoples and a better understanding of their industries.

The present publication, which is the first of two parts, is concerned mainly with the Pleistocene stratigraphy, paleoenvironments and paleogeography in and around the Douara Cave. Archeological descriptions and analyses will appear in Part II, which is scheduled to be published in 1979.

The surveys of the 1974 field season were supported by a number of sources, both individual and institutional. First of all, we should like to express our sincere gratitude to Dr. A. Bahnassi, Director of the Directorate General of Antiquities and Museums in Damascus, who gave every possible support and assistance to our expedition and kindly permitted us to borrow a complete set of materials collected from the Douara Cave for study in Tokyo.

We would also like to extend our sincere thanks to Professor A. Bounni, Director of Excavation, Dr. R. Naffakh, Head of the Prehistoric Research Section, Mr. K. Toueir, and other staff-members of the Directorate General of Antiquities and Museums in

Damascus; Mr. K. Assa'd, Director, Mr. A. Taha, Mr. K. Khamber and other members of the Palmyra Museum, who supported our excavations and geographical surveys directly in and around Palmyra; and Mr. M. Imai, the Japanese Ambassador to Syria, and the staff of the Embassy of Japan in Damascus, who kindly helped us officially and privately during our stay in Syria.

We are also indebted to the late Mr. S. Iguchi, the former President, and Mr. K. Iwanaga, Chief of Section of General Affairs, Mitsubishi Motors Corporation, who kindly offered us a motor vehicle for the expedition.

The expedition of the 1974 field season and the later investigations in Tokyo were financially supported by the Grant for Overseas Scientific Research for 1974 and 1975 from the Japanese Ministry of Education, to which we are profoundly grateful.

CHAPTER 2

Palmyra Pluvial Lake

—Paleogeography around the Paleolithic Site of Douara—

Yutaka Sakaguchi

Department of Geography, Faculty of Science, The University of Tokyo

2.1 INTRODUCTION

The author participated in the excavation of Douara Cave, carried out from August through November 1974, as a member of the Fifth Tokyo University Scientific Expedition to Western Asia, and investigated landforms around the cave in order to reconstruct the environments in which the ancient inhabitants of the Douara area lived.

The Douara Cave, correctly called the Douara Cave I, is located at 34°38.5′N and 38°27.5′E, about 20 km northeast of Palmyra (Tudmor), and situated at the lower part of the southern slope of Jabal ad Douara (Dawwara), which means a ring-shaped mountain in Arabic (Figs. 2. 1 and 2. 2). The elevation of the floor of the cave is 552.8 m. This value was obtained by a levelling started from the bench mark of Tell el-Munssef (Fig. 2. 2). Douara Cave was first excavated in 1970 by the Fourth Tokyo University Expedition, and various related investigations were carried out in the field and in the laboratory. Some results of the physical-geographical investigations have already been published (Suzuki and Takai, 1973).

What was the most impressive change in the landscape in the arid zone during the late Quaternary? It was the growth and decay of pluvial lakes. The pluvial lake is one of the most reliable records of climatic changes during the Quaternary in the present desert zone. It is the purpose of this paper to point out the appearance of pluvial lakes and climatic fluctuations in Palmyra region.

Palmyra, an oasis town which is a famous ruin from the Roman period, is located on the eastern foot of the Southern Palmyra Mountains, which extend from the southwest to the northeast. The main portion of Palmyra is 400 to 420 m above sea level. South of Palmyra there extends a vast playa called the Sabkhet Mouh. In this paper we call a spindle-shaped closed depression, which is surrounded by the main ranges of the Southern Palmyra Mountains, Jabal el-Abtar, and Tleilet ad-Dab'a, the Palmyra Basin. Its length and maximum width are about 70 km and 35 km, respectively. The lowest portion of the basin is less than 370 m above sea level. The lowest divide of this basin is situated at the eastern end of the basin and lies between 400 and 410 m in elevation. The Palmyra Basin is considered to have been formed by the intersection of a set of down-warpings, whose axes extend from the northeast to the southwest and from the east to the west, judging from its outline. The mountains and hills surrounding the basin mainly consist of limestone and marl of various geologic ages: Jurassic, Cretaceous, and Tertiary (Fig. 2. 3).

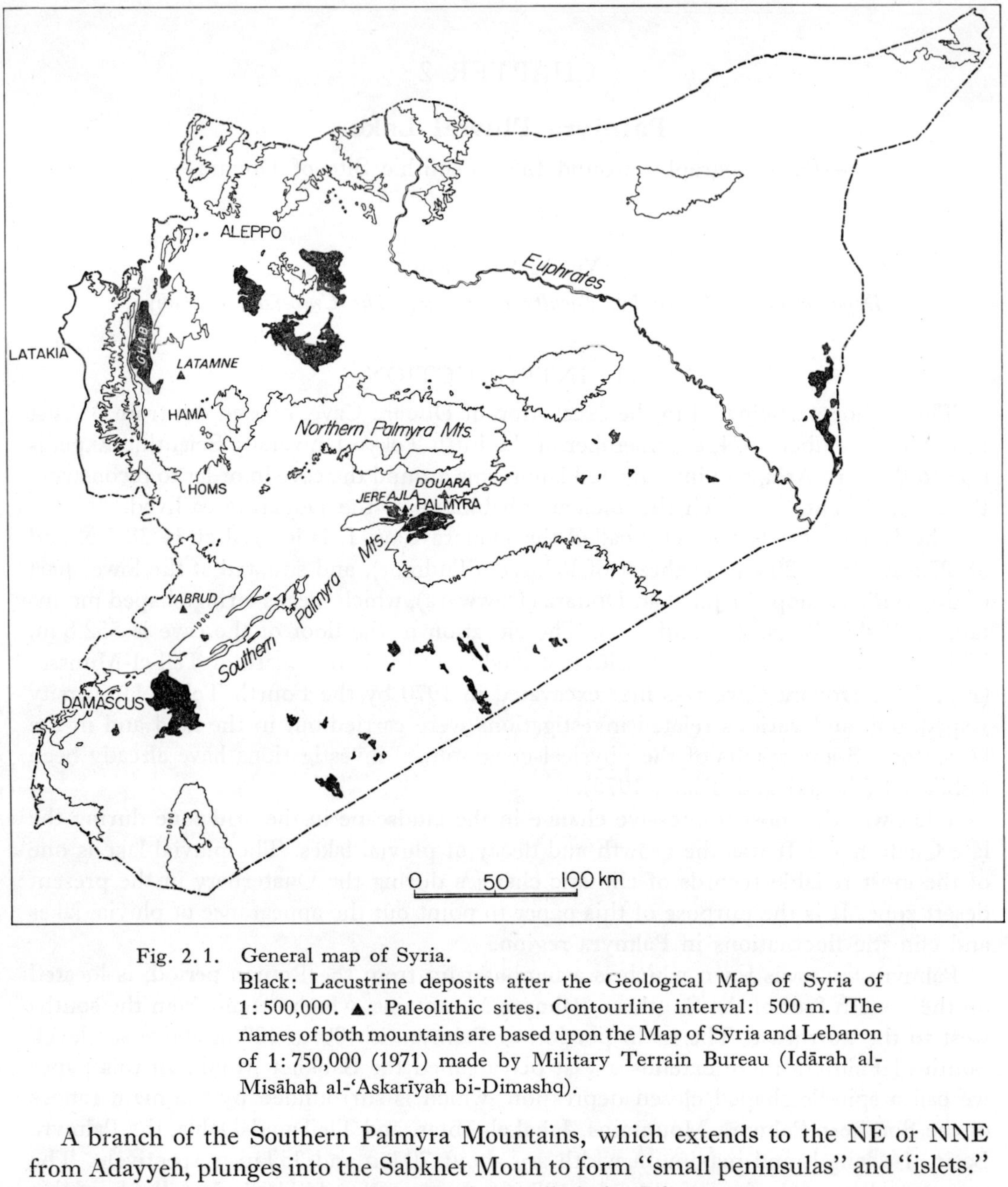

Fig. 2. 1. General map of Syria.
Black: Lacustrine deposits after the Geological Map of Syria of 1:500,000. ▲: Paleolithic sites. Contourline interval: 500 m. The names of both mountains are based upon the Map of Syria and Lebanon of 1:750,000 (1971) made by Military Terrain Bureau (Idārah al-Misāhah al-'Askarīyah bi-Dimashq).

A branch of the Southern Palmyra Mountains, which extends to the NE or NNE from Adayyeh, plunges into the Sabkhet Mouh to form "small peninsulas" and "islets." In this paper, this area east of Adayyeh is called the Islets area.

The warmest and the coldest months at Palmyra are August and January. Their average monthly temperatures during the period from 1955 to 1966 were 30°C and 7°C, respectively (Wirth, 1971). The average annual precipitation is 125 mm, but it fluctuates conspicuously from year to year. The maximum and minimum values in the nine years from 1966 through 1974 were 269 mm in 1974 and 66 mm in 1973 (Table 2. 1). The vegetational landscape of the basin is semi-desert. However, because of the extraordi-

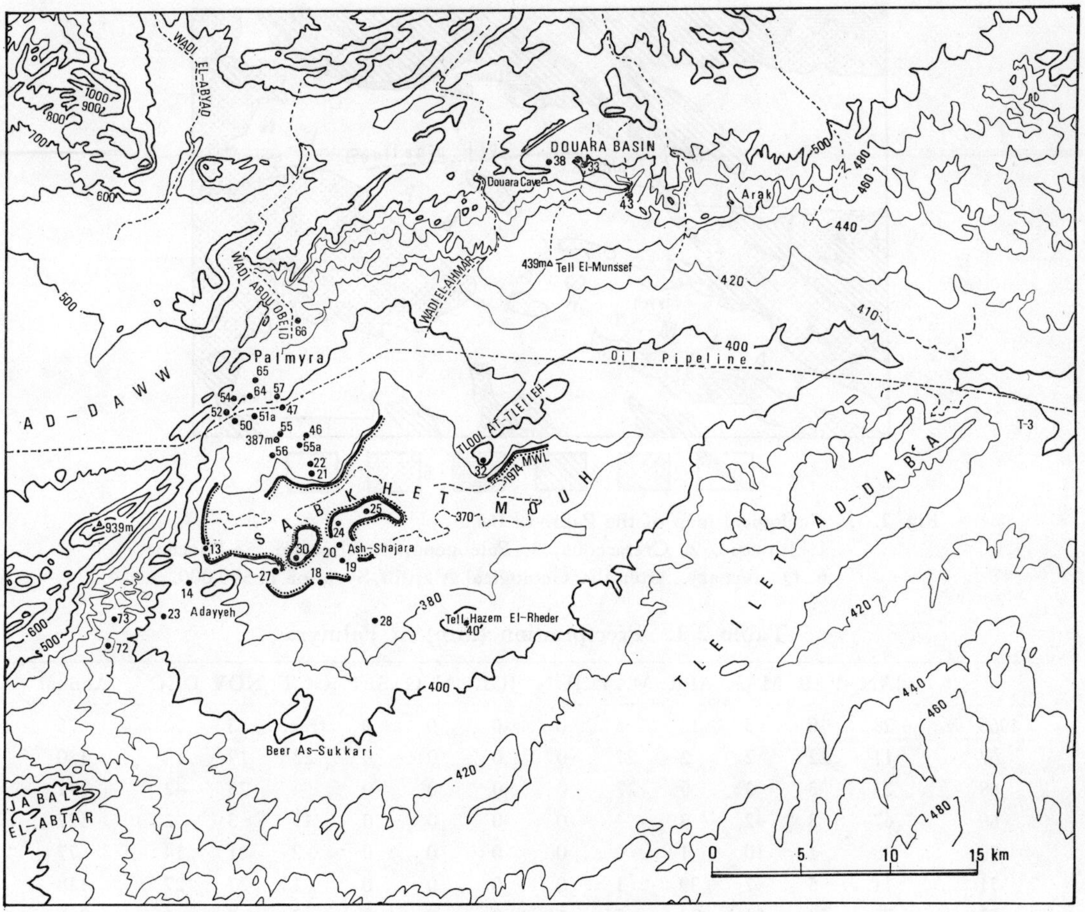

Fig. 2. 2. Topographic map of the Palmyra Basin.
1974 MWL: Maximum waterlevel of 1974 playa lake.

narily rainy and snowy winter of 1973/74 (Table 2. 2), we could still find green vegetation everywhere in the basin, though our excavation took place in the dry season.

In the Sabkhet Mouh, there are deposits of gypsum and halite. Halite deposits occur especially around Ash-Shajara (Locality 19, et al.). The Sabkhet Mouh is surrounded by pediments and alluvial fans, whose lower parts gradually change to terraces. The zonal flat between the terrace cliffs and the playa is densely covered by vegetated, moundlike sand dunes, called nebkhas (Fig. 2. 4, Plate 2. 1).

There are no perennial superficial flows in the basin. The wadis from the Southern Palmyra Mountains are relatively short but active in fluvial processes compared with those from the southern hill areas.

Lacustrine deposits are widely distributed throughout Syria, as shown in Fig. 2. 1 (Ponikarov, 1967). Though some discussions have been made concerning pluvial lakes in the Damascus Basin (Ponikarov, 1967; Kaiser et al., 1973), there is no report on the Palmyra Basin. The field surveys of this season were limited to the western half of the

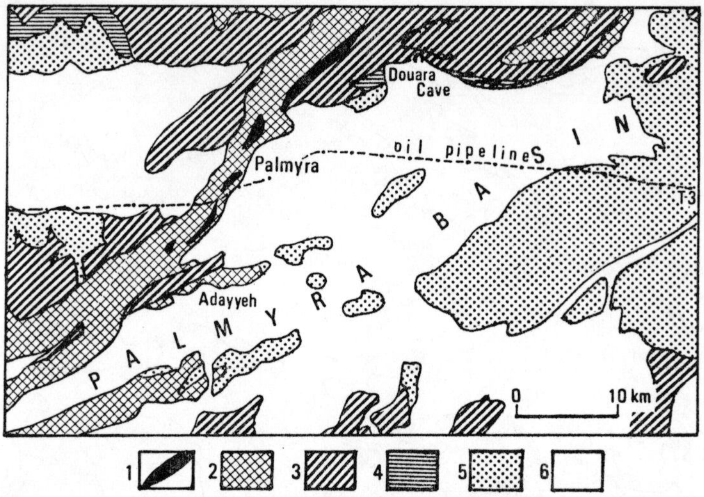

Fig. 2. 3. Geological map of the Palmyra Basin.
1. Jurassic, 2. Cretaceous, 3. Paleogene, 4. Miocene, 5. Pliocene,
6. Quaternary. After the Geological Map of Syria of 1: 500,000.

Table 2. 1. Precipitation (mm) of Palmyra

	JAN	FEB	MAR	APR	MAY	JUN	JUL	AUG	SEP	OCT	NOV	DEC	Annual
1966	28	10	5	12	5	0	0	0	0	trace	3	12	75
67	11	22	2	2	27	0	0	0	0	22	19	15	120
68	21	3	3	6	27	0	0	0	0	4	25	42	131
69	67	3	42	3	4	0	0	0	0	13	3	2	127
70	28	1	10	1	0	0	0	0	0	2	21	14	77
71	11	5	7	39	1	0	0	0	0	1	27	27	118
72	20	39	64	74	5	1	0	0	0	2	7	4	216
73	5	1	10	0	0	0	0	0	0	39	16	5	66
74	49	62	95	4	0	0	0	0	0	0	27	32	269

From Monthly Climatic Data for the World, Dept. of Commerce, USA

Table 2. 2. Precipitation (mm) of Each Rainy Season of Palmyra

1966/67	79	70/71	100
67/68	116	71/72	258
68/69	190	72/73	29
69/70	58	73/74	270

basin because of a shortage of time and because the area east of Palmyra is off-limits.

2. 2 SABKHET MOUH

Sabkhet Mouh was filled with water in 1974, because the precipitation in the winter of 1973/74 was extremely abundant, as previously mentioned. The water level of the playa lake had gone down considerably due to evaporation at the end of August, when

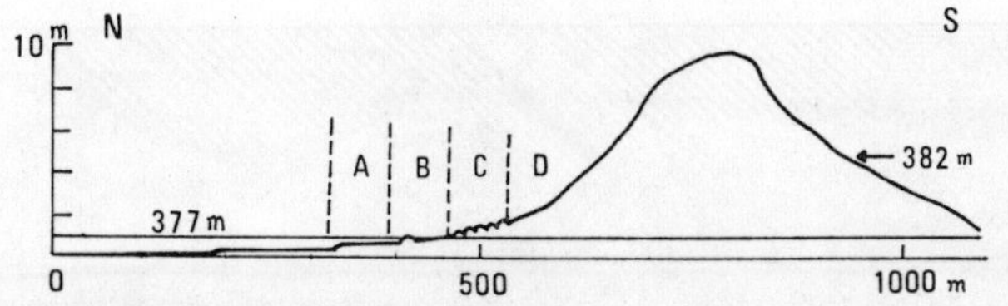

Fig. 2.4. NS-Profile through Locality 24.
A: *Arthrocnemum glaucum* polsters and algae, B: Eroded nebkhas, C: Nebkhas with *Arthrocnemum glaucum*, D: Marl covered by stone pavement, 382 m: Knick point.

Plate 2.1. Nebkha zone near Tell el-Qusseibeh. Giant nebkhas in a distant view.

we arrived there. Though the lake had nearly disappeared in the portion west of Ash-Shajara by the end of October, water still remained in the eastern Sabkhet Mouh (Plate 2.2).

The maximum water level of this playa lake was easily determined by plant remains and dung of sheep or camels, which had accumulated along the shoreline by waves, and its elevation was determined to be 377 m above sea level by a levelling based on the bench mark (387 m) south of Palmyra (Figs. 2.2 and 2.13). Parallel stripes of small wave-built terraces consisting of coarse sand and pebble remained along the maximum shoreline (Plate 2.3). A low cliff less than 1 m high had been formed by wave action in the nebkha zone between Tell el-Qusseibeh and Ash-Shajara.

The deposits of the playa lake of 1974 consisted of brown, blackish gray, or reddish brown sticky mud 5 cm thick, overlying compactly deposited sand. The dried mud surface was cracked and partly covered with algae or salt. Festucoideae polsters, which had grown on the playa surface before the water stagnated, were covered by sheet-like algal assemblages which were coated with mud particles on the outer side (Plate 2.4). In places, the nebkhas had been completely eroded by the waves and their traces remained as white patches on the dark colored ground.

According to the observation of the pit wall, dug near the maximum water level of

Plate 2. 2. Playa lake of 1974 viewed from Tlool at-Tleileh. Photographed on Sept. 17, 1974.

Plate 2. 3. Stereo pair of the lacustrine terrace at Ash-Shajara and the shoreline at the maximum water level of 1974.

1974, at Locality 21, south of Palmyra, the sand bed below the mud consisted of finer grains and included enriched horizons of gypsum crystals 1 to 7 cm thick at intervals of 3 to 30 cm (Fig. 2. 5). The gypsum crystals were usually several millimeters in size, but in places were more than 3 cm. Such a pure mud layer, as deposited in 1974, was not found in this profile. It is not clear whether such a mud layer was completely eroded before the appearance of the following playa lake, or whether it was not deposited at all until recently.

The groundwater table at Locality 21 was 103 cm below the surface on Sept. 4, 1974.

Plate 2. 4. Festucoideae polster covered with algae near Locality 13.

According to a farmer, the water table of the well in his date-palm garden, situated on a terrace north of Locality 21, was 21 m below the surface, that is, 2 m higher than usual.

A nebkha is a sand dune whose leeward slope is covered with vegetation. Neal and Motts (1967) have called such a sand dune in the western United States a phreatophyte mound. The nebkhas of Sabkhet Mouh are usually oval in slope. Their length ranges from several tens of centimeters to ten meters, and their height ranges from several tens of centimeters to five meters. The flank of a nebkha rises sharply to the surrounding plain, unlike an ordinary sand dune. Its vegetated leeward slope is particularly steep. In this case, vegetation plays an important role in obstructing drifting sand and helping a nebkha to grow; it is not a proof of its fixation. The nebkhas occur in groups. From a distance, they look like the backs of a flock of sheep. The nebkhas usually extend from the east to the west or from the northwest to the southeast. However, some nebkhas extend from the north to the south and incline steeply to the west.

Giant nebkhas about 10 m long and about 6 m high are scattered near Tell el-Qusseibeh (Plates 2. 1 and 2. 5). They extend from the east to the west or from the northwest to the southeast like ordinary nebkhas. It is said that such giant nebkhas were not be seen several decades ago.

Generally, the vegetation on the steep slopes of the nebkhas is closely related with the developmental stages. In Sabkhet Mouh *Arthrocnemum glaucum* and *Frankenia pulverulenta* (?) characteristically appear on nebkhas less than 30 cm high. The former grows exclusively on nebkhas 50 to 100 cm high. Meanwhile, *Tamarix* grows exclusively on the giant nebkhas. It is said that a period of 25 to 30 years is necessary for *Tamarix* to grow up to 2 m.

Vegetation is poor or non-existent on the gentle windward slopes of the nebkhas. A thin crust 2 cm thick with rain prints often forms on the surface (Plate 2. 1), and in places is covered by fresh eolian sand. Such a crust seems to be formed year to year by vigorous evaporation during a dry season. Such annual crusts are exposed on the eroded flank of the giant nebkhas (Plate 2. 5). The columnar profile of Locality 22 in Fig. 2. 5 shows the inner structure of a nebkha. A is fresh eolian sand. B is the crust,

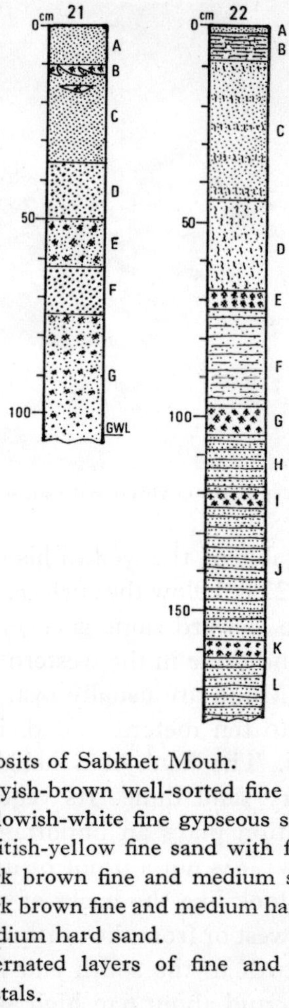

Fig. 2. 5. Recent deposits of Sabkhet Mouh.
21-A: Grayish-brown well-sorted fine sand.
B: Yellowish-white fine gypseous sand.
C: Whitish-yellow fine sand with fine gypseous sand lenses.
D: Dark brown fine and medium sand.
E: Dark brown fine and medium hard sand with gypsum crystals.
F: Medium hard sand.
G: Alternated layers of fine and medium sand and gypsum crystals.
GWL: Groundwater level.
22-A: Yellowish-gray well-sorted fine sand.
B: Yellowish-gray hard sand.
C: Yellowish-gray to grayish-yellow fine sand with plant remains.
D: Grayish-yellow fine sand with roots.
E, G, I, and K: Gypsum crystals.
F, H, J, and L: Grayish-yellow hard clayey sand.

formed definitely in 1974. C is eolian sand, in which remains of *Arthrocnemum glaucum* are buried at intervals of 5 to 8 cm. E is correlated with B of Locality 21. As the eolian sand between two successive horizons of plant remains is considered to have accumulated in a year, the nebkha is estimated to have been formed during the preceding 10 years or so. Therefore, it is concluded that the maximum shoreline of the playa lake in Sabkhet Mouh retreated from Locality 22 to Locality 21, that is, 500 m, during this time.

Wind direction in the Palmyra Basin during the excavation varied according to the locality. The dominant direction at the front of Douara Cave was SW to WSW in the

Plate 2. 5. Giant nebkha with *Tamarix* near Tell el-Qusseibeh.

Plate 2. 6. Sand dune climbing up on the mountain slope at Locality 54.

day and NE to ENE in the night from the end of August to the end of October 1974. However, the easterly or southerly winds are strong and dusty near Douara Cave. Such strong winds are effective in forming nebkhas. The wind directions observed near Douara Cave coincided with those estimated from some features of the nebkhas. Also, a snake-like sand dune ridge climbs up in a NE direction along the southern flank of the Southern Palmyra Mountains west of Palmyra (Locality 54; Plate 2. 6). This sand dune is actively moving, and seems to be very young. The dune-forming wind is also southerly. We can see several patched sand fields, which climb up the mountain slopes behind the ruin of Palmyra and shine blond-colored in the sun.

2. 3 EVIDENCE OF THE PLUVIAL LAKE

2. 3. 1 Terrace topography and terrace deposits

As previously mentioned, terrace topography is recognized everywhere around

Sabkhet Mouh, being especially conspicuous at Palmyra, Adayyeh, the Islets area, Al-Qleib, and Tlool at-Tleileh.

a) Adayyeh area

Locality 14. There are three terraces, 5.6 m, 2.0 m, and 0.6 m higher than the adjacent wadi floor. The lowest terrace has no deposits. The deposits of the middle terrace consist of gravel and yellowish-white fine sand with gypsum crystals, overlying yellowish-white hard laminated silt. The total thickness is more than 50 cm. This terrace is covered by a stone pavement, which consists of angular to subangular quartzite gravel with a maximum diameter of 10 cm. The aerated surface of this gravel is coated with desert varnish. The deposits of the upper terrace are angular to subangular gravel with a maximum diameter of 20 cm, fine sand, and silt. The gravel mainly consists of limestone, rarely flint. No quartzite gravel exists on the upper terrace.

Locality 23. The elevation of this terrace is probably 390 to 400 m. A clear terrace cliff 8 m high is formed at the eastern end. The surface is covered by a stone pavement, consisting of rounded to angular limestone and flint gravel coated with desert varnish. The terrace deposits consist of light grayish brown to light yellowish gray silt with gravel 4 m thick, lying on weathered marl. The uppermost part of the deposits is indurated. These deposits gradually change to a gravel bed lying on conglomerates or sandstone towards the mountains.

b) Islets area (Fig. 2. 6)

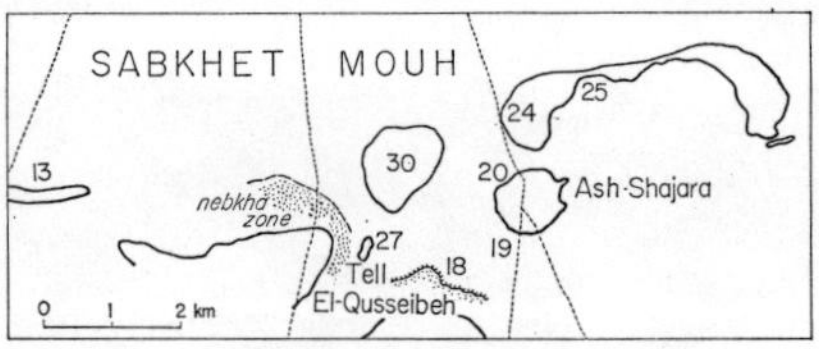

Fig. 2. 6. Sketch map of the Islets area.

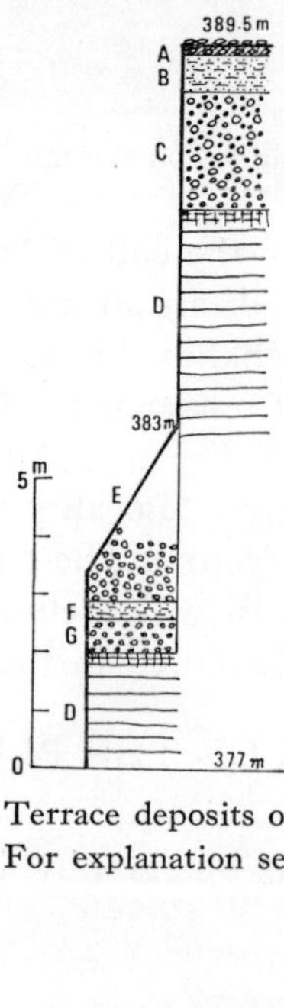

Fig. 2. 7. Terrace deposits of Locality 27.
For explanation see text.

Locality 27 (Tell el-Qusseibeh; Fig. 2. 7). This tell (or tall, small hill in Arabic) is the smallest among the isolated hills in the Islets area. The tell is flat-topped and surrounded by a steep slope. Two terrace surfaces are recognized at 12.5 m (389.5 m a. s. l.) and 6.0 m (383 m a. s. l.) above the maximum water level of 1974. The stone pavement of the upper terrace mainly consists of flint gravel and rounded flat limestone gravel with ammonite fossils with a maximum diameter of 20 cm. The deposits of the upper and the lower terraces are 2.8 m and more than 4 m thick, respectively, and their features are as follows:

A 15 cm, light brownish white crust including gypsum crystals and well-rounded flat gravel with a maximum diameter of 10 cm.

B 65 cm, light grayish-yellow sand and silt with gravel. Crumb structure.

C 200 cm, subangular to subrounded flint or limestone gravel with a maximum diameter of 10 cm.

D bed rock, yellow friable marl. The uppermost part, 30 cm thick, is weathered and includes a gypsum crystal band.

E more than 110 cm, yellowish-white gypseous fine sand and silt and gravel with a maximum diameter of 10 cm.

F 30 cm, yellowish-white fine sand and silt with gypsum crystals. Crumb structure.

G 60 cm, gravel.

Locality 30. This tell is situated 1 km north of Tell el-Qusseibeh. There are also two terraces 10 m and 6 m above the maximum water level of 1974. Their absolute heights are 387 m and 383 m, respectively. Both terrace deposits are yellowish-white gypseous sand with gravel, and are covered by the stone pavement.

Locality 25 (Plate 2. 3). Two terraces are recognized 9 m and 5 m high above the maximum water level of 1974. Their deposits are similar to Locality 27. The deposits of the upper terrace are 2 m thick, consisting of yellowish-white gypseous fine sand and silt with gravel, overlying the bed rock of marl. Some gravel is coated by Fe_2O_3 or sulphur. The terrace surface is covered by the stone pavement, consisting of angular flint gravel and rounded flat gravel with a maximum diameter of 20 cm. The surface of the lower terrace is also covered by the stone pavement, the gravel of which is about 8 cm in maximum diameter. The terrace deposits are yellowish-white gypseous fine sand and silt with gravel, more than 1 m thick.

Rounded flat gravel exists on all the above-mentioned terraces. Rounded flat gravel is common along the shore of a sea or a lake. Fine-grained sediments of fine sand or silt are also dominant in these terrace deposits. Consequently, these terraces are concluded to be lacustrine. Al-Qleib (Locality 28) is full of mesa-like flat-topped small hills 1.5 to 2 m high, consisting of light brown gypseous silt. A crust 15 to 20 cm thick has formed at the top of the deposits. No gravel is recognized here. The elevation of the hill top is estimated to be about 382 m. This area is inferred to be a dissected lacustrine plain, judging from topographic features.

The bed rocks of the isolated hills in the Islets area are marl. Limestone and flint gravels, which make the stone pavements, are exposed west of this area, near Adayyeh. This fact shows that the isolated hills and the surroundings of Adayyeh were formerly connected to each other to form a lacustrine plain. The varnish-coated quartzite gravel is only found on the lower terrace at Adayyeh. At Locality 20 of Ash-Shajara such gravel is also found on the lower terrace. The quartzite bed is exposed west of Locality 14. This bed may have been first exposed during the formative period of the lower terrace surface. The gravel may have been derived from this bed, and transported as far as Ash-Shajara in the same period.

c) Palmyra area

Locality 47. A confluent fan, formed by Wadi Abou Obeid and Wadi A'id, on which the town and the ruin of Palmyra are situated, changes to a terrace, having a steep cliff 3.5 to 4 m high at Locality 55. At Localities 46 and 55a the terrace is dissected to form some isolated hills 3 to 5.5 m high. The shallow valleys and cliffs of this area extend from the northwest to the southeast. The cause of this tendency is not yet clear, but may be related to wind erosion, judging from the dominant wind direction here. The deposits of this area are similar to those in the southwestern Sabkhet Mouh.

The vicinity of Locality 47 is the southeastern Necropolis of the Roman period (Plate

Plate 2. 7. Excavated underground tomb at Locality 47.

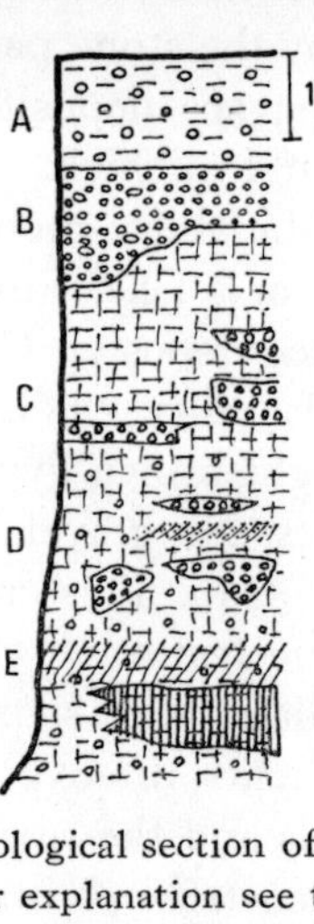

Fig. 2. 8. Geological section of Locality 47.
For explanation see text.

2. 7). The following depositional sequence is seen on the wall of an excavated underground tomb (Fig. 2. 8):

A 120 cm, yellowish-gray silt with gravel.
B 120 to 130 cm, subangular to rounded gravel with a maximum diameter of 20 cm.
C 150 cm, yellowish-gray silt with lenses of gravel.
D 20 cm, gravel.
E 250 cm, yellowish-gray hard silt with gravel, including lenses of gravel and cross-bedded sand.
F 120 cm, hard salty silt.
G 80 cm, organic clayey silt. The absolute height of the lower boundary is 389.5 m. Radiocarbon dates; 1,860±70 (TK-164) and 1,930±30 (GrN-7593).

A Middle Paleolithic flake was found in fallen earth from the nitch wall. An abraded Middle Paleolithic artifact was also found in fallen earth at the base of the cliff of Wadi

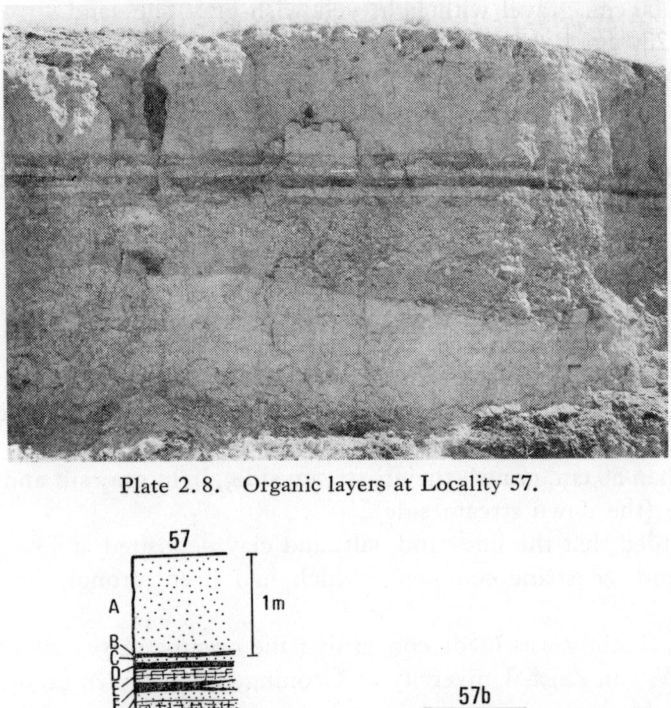

Plate 2. 8. Organic layers at Locality 57.

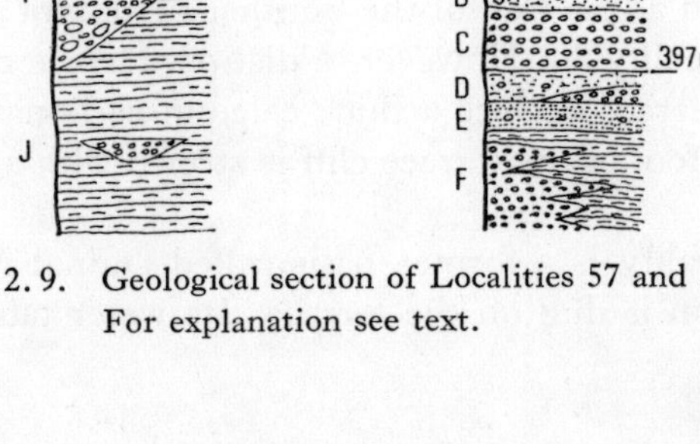

Fig. 2. 9. Geological section of Localities 57 and 57 b.
For explanation see text.

A'id at Locality 47a. Trace fossils were found in the silt layer outcropped along Wadi A'id.

Locality 57. This locality is the cliff of Wadi A'id, situated 300 m upward of Locality 47a (Plate 2. 8). The features of the deposits are as follows (Fig. 2. 9):

A 95 cm, light brownish-yellow to light brownish-gray eolian fine sand.
B 5 cm, blackish-gray organic clayey silt.
C 5 cm, light brownish-gray fine sand.
D 5 to 6 cm, blackish-gray organic clayey silt.
E 10 cm, grayish-yellow clayey silt, fine cubic structure.
F 10 cm, blackish-gray organic clayey silt. Three dark bands are seen. Radiocarbon dates; 18,900 ± 200 y.B.P. (TK-150) and 19,410 ± 150 y.B.P. (GrN-7597).
G 10 cm, light grayish-brown fine sand. Lower partly organic.
H 5 to 10 cm, light yellowish-gray silty fine sand, slightly organic, fine cubic structure.
I 50 to 100 cm, gravel with light yellowish-gray fine sand.
J 120 to 230 cm, light yellowish-gray to light brownish-gray hard massive clayey silt, including lenses of gravel, prismatic structure.

F layer is nearly horizontal between Locality 57 and 57a, 61.5 m upward of Locality 57.

At Locality 57a three terraces are recognized at heights of 1.5 m, 2.5 m, and 4.5 m. The lowest terrace disappears 365 m upward of this locality to make a knick point.

Locality 57b. This locality is situated at the cliff of Wadi A'id 290 m up Locality 57a. The depositional sequence is as follows (Fig. 2. 9):

A 50 cm, surface soils.
B 20 cm, light gray fine sand and silt with gravel.
C 40 cm, subangular to subrounded gravel with a maximum diameter of 10 cm.
D 20 cm, light gray silt and fine sand with lenses of gravel, fine cubic structure.
E 20 cm, light gray laminated fine sand with gravel.
F more than 50 cm, gravel (the up-stream side), light gray silt and clay, fine cubic structure (the down-stream side).

It was concluded that the fine sand, silt, and clay deposited at Localities 47 to 57b were telmatic and lacustrine sediments which had been strongly affected by fluvial processes.

A palynological study was made concerning the organic clayey silts at Localities 47 and 57 by Dr. W. van Zeist, University of Groningen, but pollen grains were very rare and undeterminable.

Near Localities 50 and 52.5 km southwest of Palmyra, there are terraces contiguous with the pediment formed at the foot of the Southern Palmyra Mountains. The terrace cliff is not so clear at Locality 50. However, a distinct terrace cliff appears 500 m west of this locality. On the terrace surface a thick calcium carbonate crust is formed. The 80-m-wide flat along the foot of the terrace cliff is strewn with angular gravel and many artifacts.

Locality 50. This locality is a former oasis called Ouinat Mohmed el-Fard (Plate 2. 9). A well for irrigation is dug on the terrace. Its water table is 270 cm below the

Plate 2. 9. Ouinat Mohmed el-Fard (Locality 50) viewed from the eastern side.

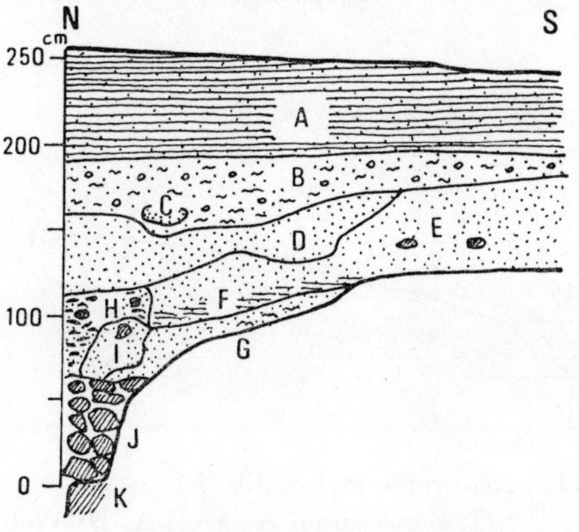

Fig. 2. 10. Geological section of Locality 50.
For explanation see text.

surface. The depositional sequence of the sediments seen on the well side is as follows (Fig. 2. 10):

A 65 cm, light reddish-brown to light grayish-brown laminated medium to fine sand.

B 30 to 40 cm, light gray mottled fine to medium sand with gravel, plentiful plant remains, sand blocks including arrtifacts (C). The radiocarbon date of the *Phoenix* bark lying at the lower boundary is 38,450±650 y.B.P. (GrN-7598).

D 40 cm, light brownish-gray fine to medium sand.

E 60 cm, light gray fine to medium sand, fragments of the crust, lower part light gray clayey fine to medium sand (F).

G 20 cm, light gray medium to fine sand with plant remains.

H 50 cm, light gray fine to medium sand, many artifacts and some animal bones.

I 30 cm, dark gray hard medium to fine sand.

J 70 cm, blocks of the crust.

K calcium carbonate crust.

The artifacts collected from these deposits belong to the Kebaran and the Neolithic ages. On the terrace immediately north and east of this well many artifacts which belong to the Kebaran and the Geometric Kebaran ages, were collected. The elevation of the top of the crust (K) is 382 m.

Locality 52. There are two wells, called A and B, immediately north of the Homs-Palmyra road. Well B has collapsed, but well A is now used for irrigation of the palm garden, situated south of the road. A calcium carbonate crust has formed all over the surface between the road and a knick point at the foot of the mountains shown in Fig. 2. 11 (Plate 2. 10). The crust in this locality consists of light gray fine sand indurated by calcium carbonate. The crust is so hard that a hammer is necessary to pick out a piece. The crust usually has a platy structure and in part, moreover, a honeycomb structure. A number of artifacts of the Epi-Paleolithic, Pre-Pottery Neolithic A and B stages and of the Middle Paleolithic type are found in the crust around well B.

Well A is 5 m in diameter. Its water table is 5 m deep and mostly covered with reeds. The boundary between the terrace deposits and the bed rock outcrops is immediately above the water table. The groundwater springs out from two outlets on the boundary.

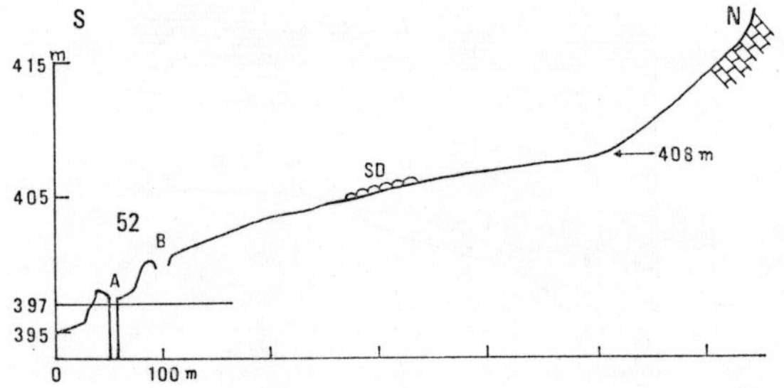

Fig. 2. 11. NS-profile at Locality 52.
SD: Sand dune, A: Well A, B: Well B.

Plate 2. 10. Calcium carbonate crust near Locality 52.

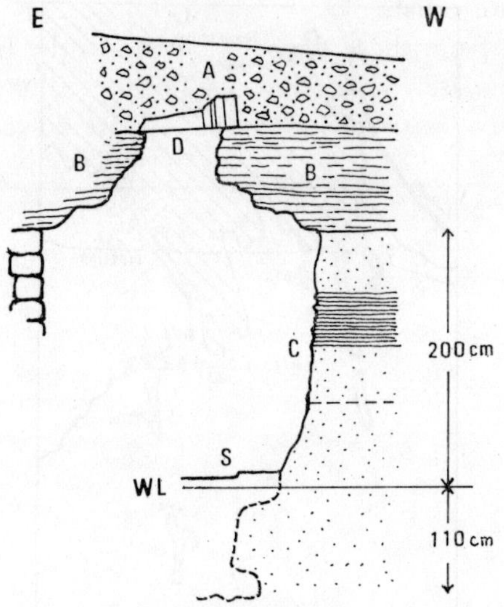

Fig. 2. 12. Geological section of Locality 52.
WL: Water level, S: Slice.

Well A is connected by a tunnel with the palm garden. The depositional sequence on the wall of the tunnel is as follows (Fig. 2. 12):

A 150 cm, light gray fine sand with gravel, artificial.

B 60 cm, light gray to light brownish-gray encrusted sand containing artifacts. D is a piece of treated stone of the Roman period.

C more than 200 cm, light gray to white salty fine sand. The middle part is remarkably salty and laminated.

C layer is inferred to be lacustrine, based on its lithologic features. This layer has no gravel, thus differing from the other localities of Palmyra. Light brownish-gray to light gray laminated fine sand 110 cm thick is lying on the crust. This sand is slightly indurated, but can be easily crushed by kicking.

2. 3. 2 Fossil river mouths

Snake-like hills extend like fingers towards the south from the Palmyra confluent fan as shown in Fig. 2. 13. Their absolute height ranges from 380 to 390 m. The hills are 1.5 km long, 50 m wide, and 5.5 m high in maximum value. They trend from N30°E to N40°W. The western hills are higher than the eastern ones in relative height. At Locality 56, the relative height decreases toward the south, that is, 2.5 m at the north end, but 1 m at the south end 500 m from the north end. These hills consist of light brownish-gray fine sand or silt including well-rounded to subangular gravel. Their surface is covered by a stone pavement. At Locality 51a flat gravel is included in the stone pavement. The maximum diameter of the gravel is 20 cm at Locality 51. At Locality 55 the gravel size decreases toward the south, that is, 15 cm at the north end, and 8 cm at the south end 490 m from the north end. At Localities 51 and 51a, a terrace is formed at the middle of the hill slope. The hill slope at Locality 51a is slightly dissected by shallow valleys. At Locality 51b the eroded hill remains as a flat sand field.

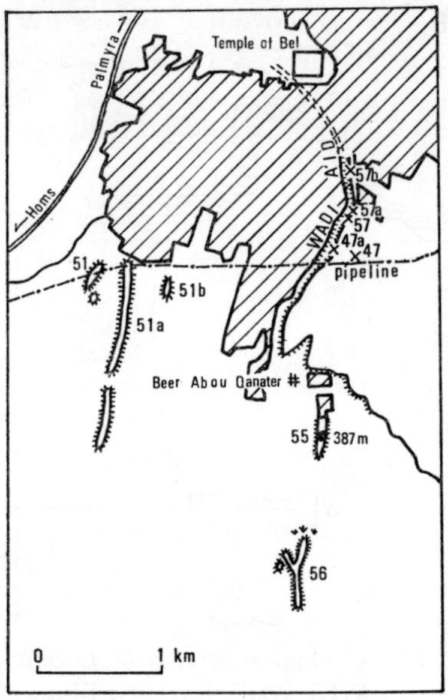

Fig. 2. 13. Former river mouths south of Palmyra.
Shadowed portion: Palm garden.

These hills are former distributary river beds, which extended into a lake like the typical bird's foot delta of the Mississippi River. A number of Middle Paleolithic and post-Middle Paleolithic artifacts are found on the hills. The former are abraded, but the latter are fresh. The river mouths are inferred to have stopped their growth in the Middle Paleolithic age or soon after its end. The end of growth may have been caused by a decrease in fluvial action and/or the disappearance of a lake due to climatic changes.

2. 3. 3 *Conclusion*

As previously mentioned, the landforms and the deposits of Sabkhet Mouh and its surroundings coincide with each other on a lacustrine origin. A diatom analysis was tentatively carried out on some of the samples collected from the main localities; however, no diatom was found in them. Nevertheless, it is certain that a pluvial lake existed at least in the Paleolithic ages in the Palmyra Basin. The absence of diatom remains may be explained by dissolution of the diatom frustules after death, because of the high pH value of the pluvial lake, like Lake Victoria (Kendall, 1969). The lake is called the Paleo-Palmyra Lake.

2. 4 HISTORY OF THE LAKE AND CLIMATIC CHANGES

Judging from the geological data of Locality 57, the lake existed about 19,000 years ago. It is concluded that its water level lay at 397 m above sea level, as the boundary between the C layer of fluvial facies and the D layer of lacustrine facies lies at the same level.

Also, the organic clayey silt at Locality 47 lies 5 m lower than the organic layers at Locality 57. Both of the radiocarbon dates are too young. The dates coincide with the age of the underground tomb. The discordance between the stratigraphy and the dates may be caused by the following: 1) the original sequence of the deposits was disturbed when the underground tomb was constructed; 2) carbon exchanges took place between the organic matter of the deposits and bodies buried in the Roman period, because they came into contact with each other.

The author does not have sufficient data to conclude whether or not a lake existed more than 19,000 years ago. A piece of *Phoenix* bark found at the well of Locality 50 is suggestive regarding this problem. The radiocarbon date of this bark is 38,500 y.B.P. This bark was probably transported at the time of a flood and buried here. The stand of *Phoenix* is surely between Locality 50 and the mountain 1 km from this locality. For the stand of *Phoenix*, a high groundwater table seems to be suitable. Such a place may be Locality 52, where the groundwater springs out now. The present water table of this locality is very low. If the lake surface were as high as 395 to 400 m, the water table would lie near the ground surface. It is considered that the dated bark was brought from *Phoenix* growing at Locality 52 under such a condition. A lake probably also existed 38,500 years ago.

According to the palynological studies on the 11 m thick deposits of the Ghab depression, western Syria, a low percentage of non-arboreal pollen appears at the pollen assemblage zone (T) 8.5 to 9.5 m deep. The age of this zone is inferred to be more than 45,000 years. J. Niklewski and W. van Zeist (1970) concluded that at that time the climate would have been at least as moist as it is today, and temperatures would have been lower than at present. The birth of the Paleo-Palmyra Lake could go back to this time. In the Damascus Basin pluvial lakes were formed twice during the Quaternary, that is, in the Riss and Würm ages. However, a pluvial lake in the Riss age is not yet confirmed in the Palmyra Basin.

The water level of the lake 38,500 y.B.P. is estimated to have been about 395 to 400 m above sea level. According to the topographic map, the elevation of the lowest divide of the Palmyra Basin is between 400 and 410 m. Therefore, the water level of the lake cannot exceed this elevation. This fact is consistent with the geological data on Sabkhet Mouh and its surroundings. However, if the lowest shallow and wide divide facing T-3 resulted from erosion or downwarping, one can hypothesize that the maximum lake level was higher than 400 m in the Palmyra Basin. At Locality 72 the remarkable terraces are seen at more than 400 m above sea level (Fig. 2. 14; Plate 2. 11). Though these terraces seem to be the key to the solution of the higher water level problem, the author

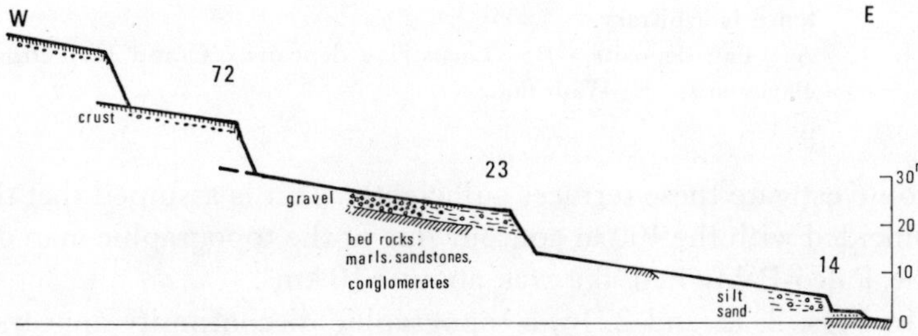

Fig. 2. 14. Schematic EW-profile at Adayyeh and its surroundings. Horizontal distance is arbitrary.

Plate 2. 11. Terrace with superficial crust at Locality 72.

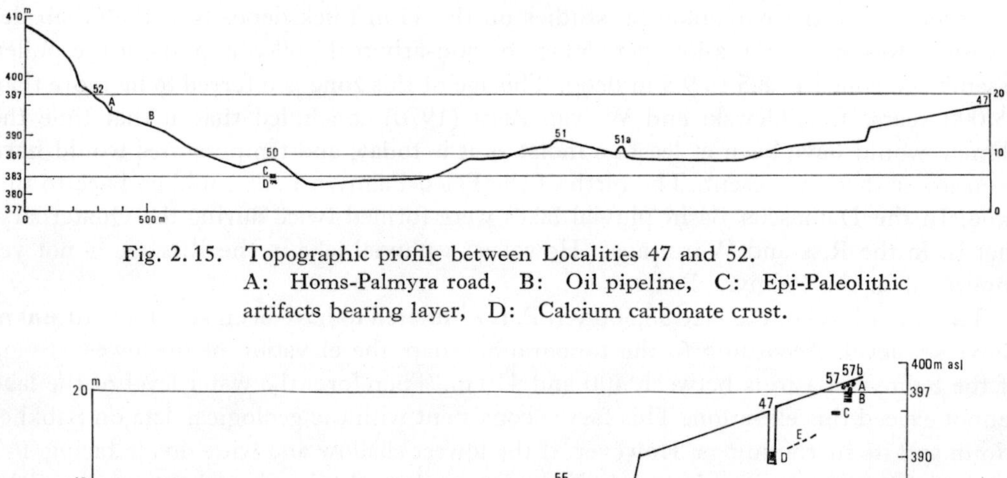

Fig. 2. 15. Topographic profile between Localities 47 and 52.
A: Homs-Palmyra road, B: Oil pipeline, C: Epi-Paleolithic artifacts bearing layer, D: Calcium carbonate crust.

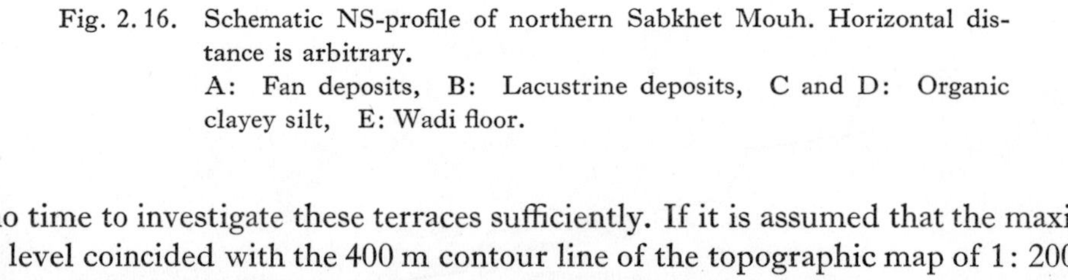

Fig. 2. 16. Schematic NS-profile of northern Sabkhet Mouh. Horizontal distance is arbitrary.
A: Fan deposits, B: Lacustrine deposits, C and D: Organic clayey silt, E: Wadi floor.

had no time to investigate these terraces sufficiently. If it is assumed that the maximum water level coincided with the 400 m contour line of the topographic map of 1: 200,000, the area of the Paleo-Palmyra Lake was about 510 km².

As shown in Figs. 2. 15 and 2. 16, a topographic discontinuity appears at 387 m or so around Palmyra. This shows a stagnation of the water level during the period of decline of the Paleo-Palmyra Lake. The 387 m terraces are common around the western

Table 2. 3. Maximum Diameter of Terrace Gravels

Locality	Middle terraces	Higher terraces
Tell el-Qusseibeh (27)	10 cm	15 cm
Qusseibeh (30)	7	25
Ash-Shajara (25)	8	20
Beer Abou Qanater	10	20
Tell Hatem el-Rheder (40)	10	20
Wadi Abou Obeid (66)	25	50
Douara Basin (33)	40	50
Douara Basin (38)	20	30
Wadi Jaria (43)	50	60

Sabkhet Mouh; Localities 24 (Ash-Shajara, 386 m), 30 (Qusseibeh, 387 m), 13 (Adayyeh, 385.5 m), and 32 (Tlool at-Tleileh, 387 m).

Another topographic discontinuity is also recognized at 383 m above sea level. In particular, this discontinuity is clearly recognized in the Islets area.

The formation of the lake was caused by the increase of precipitation in the Palmyra Basin. The terraces in this basin are divided into three groups, upper, middle, and lower. However, this division and the correlation of the terraces are only tentative. The upper group includes the terraces correlated with the 397 m and 387 m lacustrine terraces. The middle terraces include the terraces correlated with the 383 m lacustrine terraces. As shown in Table 2. 3, the grain size of the upper terrace deposits is larger than that of the middle ones. This fact shows that fluvial action was stronger in the upper terrace formation period than in the middle one. Fluvial actions are a function of discharge, which depends on precipitation in a drainage basin. Therefore, the precipitation in the Palmyra Basin is estimated to have decreased from the period of the 397 m and 387 m terraces towards the period of the 383 m terrace. The decline of the lake was caused by the decrease in precipitation, not by the increase in evaporation as a result of the rise of temperature. The period of the high water level coincides exactly with the Würm glacial age. It is concluded that the precipitation of this period was brought about in the summer by the polar front, which shifted southward (Suzuki, 1972). The precipitation was more abundant than today. As it is inferred to have been cooler in this period than it is today in the Damascus Basin and the Ghab depression, the same tendency is also expected in the Palmyra Basin. The ground in the Palmyra Basin may have been more densely covered with vegetation than at present, because of summer rainfall. The bulky hearth ash containing Middle Paleolithic artifacts found in Douara Cave in 1974 is considered as evidence that plant population in the Palmyra Basin was sufficient for fuel or food for wild animals all the year round during this period. One of the radiocarbon dates of this ash is dated more than 52,000 y.B.P. (GrN-7599). This date is favorable to the above-mentioned inference that the pluvial lake may have been born before 45,000 y.B.P.

Since 19,000 y.B.P. the polar front may have shifted northward, and thus not brought much rainfall to the Palmyra Basin. Consequently, the water level of the lake became lower discontinuously. The river mouths south of Palmyra ceased their extension during

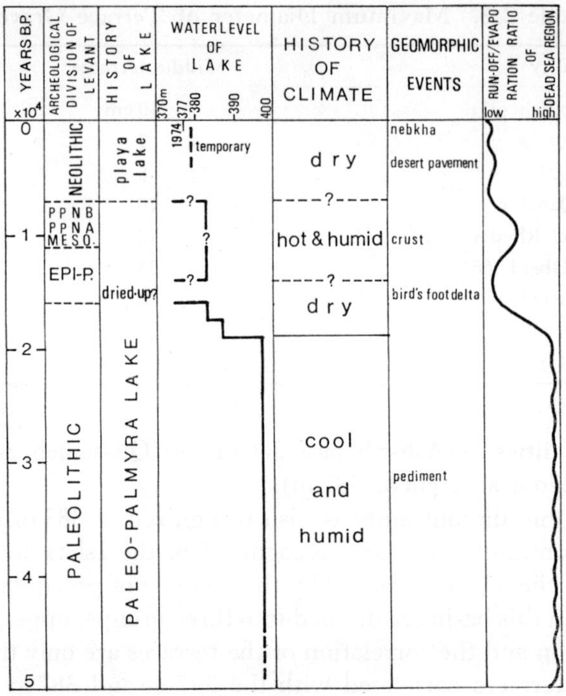

Fig. 2. 17. Paleogeography of the Palmyra Basin.
Epi-P.: Epi-Paleolithic, Meso.: Mesolithic, PPNA: Pre-Pottery Neolithic A, PPNB: Pre-Pottery Neolithic B. The water level line shows only a general tendency.

this time. In the Ghab pollen diagram, a high percentage of non-arboreal pollen is recognized 1.5 to 2.0 m deep, dated 16,000 to 14,000 y.B.P. The time of this zone is inferred to have been dry. As desiccation from 16,000 to 14,000 y.B.P. is commonly recognized in the whole of the Near East, Paleo-Palmyra Lake is inferred to have lowered its water level very much or completely dried up. At Tabūn und M. Wadi, Palestine, the *Dama* (moist) / *Gazella* (dry) ratio decreases from the excavation level of Wad D to that of Wad B (Garrod and Bate, 1937). Therefore, in Palestine the desiccation proceeded from the middle stage of the Upper Paleolithic. In the Dead Sea region the climate is also inferred to have rapidly become drier from 20,000 y.B.P. (Fig. 2. 17; Farrand, 1971). The desiccation from 19,000 y.B.P. in the Palmyra Basin coincides with these circumstances.

The calcium carbonate crust seen around Locality 52 is commonly formed everywhere from Locality 65 near Efqa Spring (Palmyra Hammam) to Locality 72 along the foot of the Southern Palmyra Mountains, and it consists of limestone and calcareous rocks. W. J. van Liere (1960/61) has concluded that the calcium carbonate crust in Palmyra was formed in the early Quaternary, because there is not sufficient calcium in the solution to precipitate with a rainfall of less than 200 mm. At Locality 72, the terrace surfaces and the upper part of their slopes are covered by a calcium carbonate crust 50 cm thick. Therefore, the formation of the crust took place after the formation of the terraces. Moreover, the crust around Palmyra includes Epi-Paleolithic artifacts. On the other hand, the surface soils bearing the Kebaran and the Geometric Kebaran artifacts at Locality 50 are not encrusted. However, the stratigraphical relationship between the well and the archeologic sites around the well at Locality 50 is not sufficiently clear. According to the Ghab pollen diagram, it is inferred to have been somewhat moister in the period between 10,000 and 8,000 y.B.P. This period seems to be suitable for the formation of a crust. Though we have not yet found artifacts more recent than Epi-Paleolithic in the crusts near Palmyra and cannot confirm the strict time of the crust formation, it is certain that the crust was formed at a time between the Epi-Paleolithic and the Pre-Pottery Neolithic.

The crust may have been formed in the following manner: rain water, deeply infiltrated into the ground, rises again up to the ground surface by capillary action and evapotranspiration, and calcium carbonate precipitates in the interstice of the soil. A climate in which rain mainly falls in the hot season is suitable for this process. Accordingly, the rainy season of this time in the Palmyra Basin was the summer. The temperature of the summer half year is implied to have been as high as at present. Sabkhet Mouh may have been filled with water. Its maximum level is considered to have been no more than about 383 m above sea level.

At the ruin of Palmyra many columns have been abraded at the bottom by sand blasting so completely that their flutings have disappeared. Judging from this fact, it is concluded that the climate was dominantly dry since the third or the sixth century, when the city wall around the ruin was destroyed. A minor topographic discontinuity is recognized at about 379 m above sea level. This level coincides with the upper limit of the nebkha zone. This knick point may have been formed at the highest water level of the playa lakes. The giant nebkhas are said to have been formed during the last few decades. The nebkha zone is certainly extending. The sand dune at Locality 54 is actively climbing up the mountain slope. The diminishing of the playa lake area suggests a decrease in rainfall. It is said that the Syrian Desert supported large herds of gazelle and Syrian wild ass (*Equus hemionus*) as recently as 200 years ago (Schroeder, 1970). *Tamarix* which formerly grew in Sabkhet Mouh was felled for use as material for fuel and dwelling. The desertification around Palmyra is considered to have been brought about not only by climatic fluctuations, but also by man's impact on the vegetation.

Acknowledgement

The author wishes to thank Dr. W. van Zeist, Institute for Biological Archaeology, University of Groningen for making palynological studies, and Prof. H. Kobayashi, Tokyo Gakugei University for making a diatom analysis.

LITERATURE CITED

Farrand, W. R. (1971) Late Quaternary Paleoclimates of the Eastern Mediterranean Area. In: Turekian, K. K. (ed.), The Late Cenozoic Glacial Ages. xii+606 pp., Yale Univ. Press, New Haven, 529–564.

Garrod, D. A. E. and Bate, D. M. A. (1937) The Stone Age of Mount Carmel, I. 240 pp., Oxford at the Claredon Press.

Kaiser, K., Kempf, E. K., Leroi-Gourhan, Arl., und Schütt, H. (1973) Quartärstratigraphische Untersuchungen aus dem Damaskus-Becken und seiner Umgebung. Z. Geomorph. N. F., **17**: 263–353.

Kendall, R. L. (1969) An Ecological History of the Lake Victoria Basin. Ecol. Monogr., **39**: 121–176.

van Liere, W. J. (1960–61) Observations on the Quaternary of Syria. Berichten Rijksd. Oudheidk. Bodemonderzoek, **10–11**: 7–69.

Neal, J. T. and Motts, W. S. (1967) Recent Geomorphic Changes in Playas of Western United States. J. Geol., **75**: 511–525.

Niklewski, J. and van Zeist, W. (1970) A Late Quaternary Pollen Diagram from Northwestern Syria. Act. Bot. Neerl., **19**: 737–754.

Ponikarov, V. P. (ed.) (1967) The Geology of Syria. Explanatory Notes on the Geological Map of Syria, I. Ministry of Industry, Syrian Arab Republic, 229 pp.

Schroeder, H. B. (1970) The Paleolithic Industries from the Syrian Desert Cave of Jerf Ajla. Etudes sur le Quaternaire dans le Monde, **2**: 1017–1024.

Suzuki, Hideo (1972) World Climates of the Würm Glacial Age. The Quaternary Research, **11**: 171–180.

Suzuki, Hisashi and Takai, F. (eds.) (1973) The Paleolithic Site at Douara Cave. Univ. Museum, Univ. Tokyo, Bull. **5**: 1–150.

Wirth, E. (1971) Syrien. xx+530 pp.+16, Wissenschaftliche Buchgesellschaft, Darmstadt.

CHAPTER 3

Climato-genetic Landforms around Jabal ad Douara and Its Surroundings

Takeei Koizumi

Department of Geography, Faculty of Science, The University of Tokyo

3.1 INTRODUCTION

In this paper the author deals with the history of the natural environment, especially of the geomorphic development and climatic change during the late Quaternary, in the Palmyra region. Mainly studied is the area between Jabal ad Douara (Douara Mountains), which is bounded on the south by the Palmyra Basin, and the playa lake in the central part of the Palmyra Basin. As the Palmyra region is located in an inland area, it is generally difficult to correlate the geomorphic surfaces in this region with those in other regions. However, many kinds of climato-genetic landforms and the flint artifacts, which correspond to the several developmental stages, give sufficient ideas on the natural history of the Palmyra region.

3.2 GEOGRAPHICAL SETTING OF THE PALMYRA REGION

The Palmyra Basin is located in the northern part of the Syrian Desert. It is a structural basin, which is bounded on the north and west by mountains at around 1,000 m

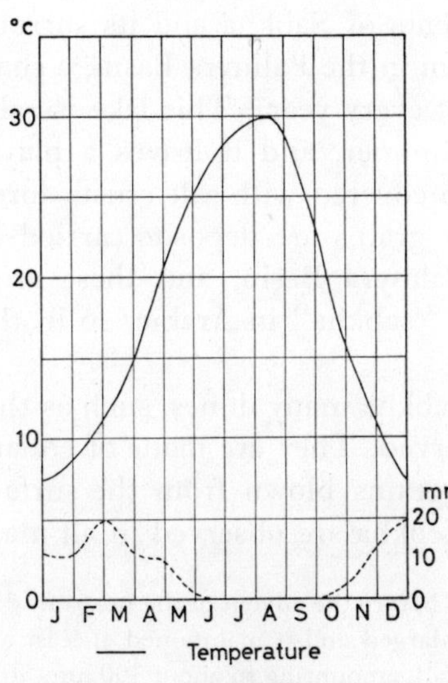

Fig. 3.1. Climate of Palmyra (after Abdul-Salan).

above sea level and on the south by a basalt plateau. On the east it is separated from the Euphrates lowland by a hilly district. The central part of this basin is at an altitude of 370 m above sea level.

There are two characteristic seasons in the Palmyra region: a dry and hot season and a wet and cold season (Fig. 3. 1). In August, the average and the maximum daily temperatures exceed 30°C and 40°C, respectively. In addition, a hot wind frequently rises. In winter, on the other hand, the average daily temperature is 7.4°C and the minimum at night is often below freezing. The annual rainfall is only 125 mm and the monthly rainfall during the short rainy period in winter is characteristic of it. During this season, there are snowfalls which sometimes amount to 50 cm. As is mentioned above, this region belongs to a semi-desert climatic area with a Mediterranean climatic character in its rainfall. Consequently, the vegetation cover is scarce in this region, though flora is rather abundant. The main families growing in the Palmyra region are *Graminea*, *Chenopodiaceae*, and *Compositae* (Zohary, 1973; Mouterde, 1970). Plants grow temporarily in early spring, immediately after the rainy season. In the other seasons, the region turns brown in color except along the wadis and a salt lake.

3. 3 PRESENT GEOMORPHIC PROCESSES

In order to restore the natural environment, especially the landforms and climate of the past, the present geomorphic processes should first of all be made clear. The author therefore paid attention to the processes from debris production to sedimentation under the present climatic conditions, and he surveyed the mountain slopes as erosional areas, the wadis as areas of removal, and the salt marsh as a sedimentary basin. It is thought that a change in the natural environment would cause a change in this whole system and that it would be reflected in the landforms.

To begin with, the landforms and sediments in the playa lake and its surroundings can be described as follows:

(1) Landforms and sediments of Sabkha and its surroundings

As there is no outflow stream in the Palmyra Basin, a shallow temporal lake appears during the rainy season almost every year.[1] This lake is salty. It dries up and shrinks during the long drought in summer, and it leaves a playa area composed of small particles (Fig. 3. 2), which are covered with salt crusts formed by salt concentration. These small particles and salt grains are deposits carried by the wadis which rise in the mountains around the Palmyra Basin, and these are very well-sorted. In the Palmyra region playa is called "sabkha" in Arabic, so in this paper the latter term is adopted.[2]

In the marginal zone of a sabkha many dunes, such as the mounds called "nebkha" (Abdul-Salam, 1965), are observed. They are made of eolian deposits, which are composed of sand, silt, and salt grains blown from the surface of the sabkha, and they are kept in place by plants. Nebkha are observed in all marginal zones of sabkha and

1. This salt lake usually vanishes before the advent of the next rainy season. In the 1974 season when we visited, however, the lake was enlarged and it maintained at least a depth of 1 m till the next rainy season because of the abnormal rainfall, amounting to about 300 mm, in the previous winter.

2. There are many terms for playa: sabkhah (Powers, et al., 1966), mamlahah (Holm, 1960) or inland sebkhas (Glennie, 1970) etc. are used in the literature.

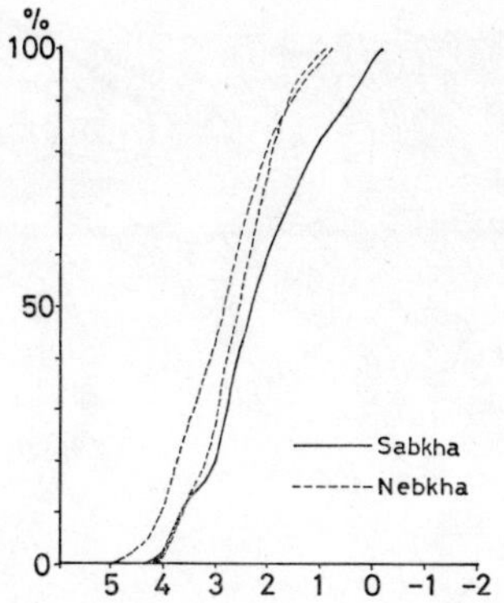

Fig. 3. 2. Cumulative curves of grain size composition of the deposits of sabkha and nebkha.

are found also along the wadi channels. They are well developed, especially on the eastern and northern margin of the sabkha, and they form a nebkha zone which is over 2 km in width. Nebkha are generally 1–3 m in length, 1–2 m in width, and 30–80 cm in height but, sometimes they are smaller. Their long axes coincide with the prevailing wind direction (nearly E-W). Halophilous plants which belong to *Halopeplis* and *Halocnemus* ofen grow on the tops of the nebkha. The nebkha are still growing. On its vertical section, cross-laminations and buried halophilous plants are observed. Judging from this section, it is known that the competition between the growth of the plants and the sedimentation of the sand grains created the nebkha landforms. The sand grains composing nebkha are well sorted as shown in Fig. 3. 2. The size distribution of its grain is similar to that of the sabkha surface deposits, which means that the sand grains were derived from the sabkha.

(2) Erosion, transportation and sedimentation by a wadi

Several wadis flow into the sabkha from the surrounding mountains. In these wadi channels, the streams exist only during a short rainy season, and they dry up in the dry season.

Plate 3. 1. Landscape of Douara Basin.

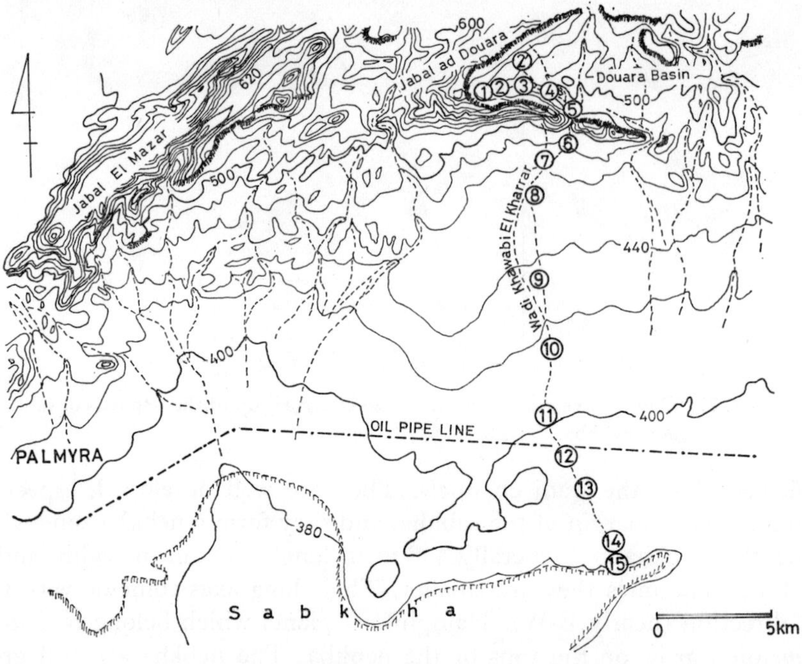

Fig. 3. 3. Sampling points of wadi deposits.

The landform formation by the wadis can be described by taking Wadi Khawabi el Kharrar, which rises in Jabal ad Douara, as an example. This wadi is about 30 km long and in the upper reaches of the stream it creates a small basin (Douara Basin,[3] Plate 3. 1) in Jabal ad Douara. It flows southwards through the northern part of the Palmyra Basin to sabkha. The channel deposits were sampled to examine transportation by the wadi at several points along the wadi, as shown in Fig. 3. 3. The grain size distribution is shown in Fig. 3. 4.

The Douara Basin is a chestnut shaped basin. Its diameter is about 8 km E-W and 5 km N-S. This basin originates from the anticlinal valley eroded by Wadi Jaria which comes from the southeast. The mountains, which restrict the southern margin of this small basin, are eroded to several interrupted peaks and there are some cols between these peaks. Wadi Khawabi el Kharrar divides into two main branches at the entrance of Douara Basin, which has many tributaries flowing into it from the mountain fronts. Along the mountain front big knick points are formed and several dry falls are found.

The Douara Basin is enclosed by mountains (200 m in relative height) that consist of Cretaceous and Paleogene muddy limestone (Wolfart, 1967). The mountain slopes have high cliffs, talus slopes and concave slopes which are situated between the cliffs and the bottom of the basin. At present the debris supply from the mountain slopes is poor, and the wadis are cutting the talus slopes or the piedmont gentle slopes, and older geomorphic surfaces are terraced. At point 1 in Fig. 3. 3, the wadi cuts the basal

3. "Douara" means "ring" in Arabic. It is derived from the form of the mountains surrounding this basin.

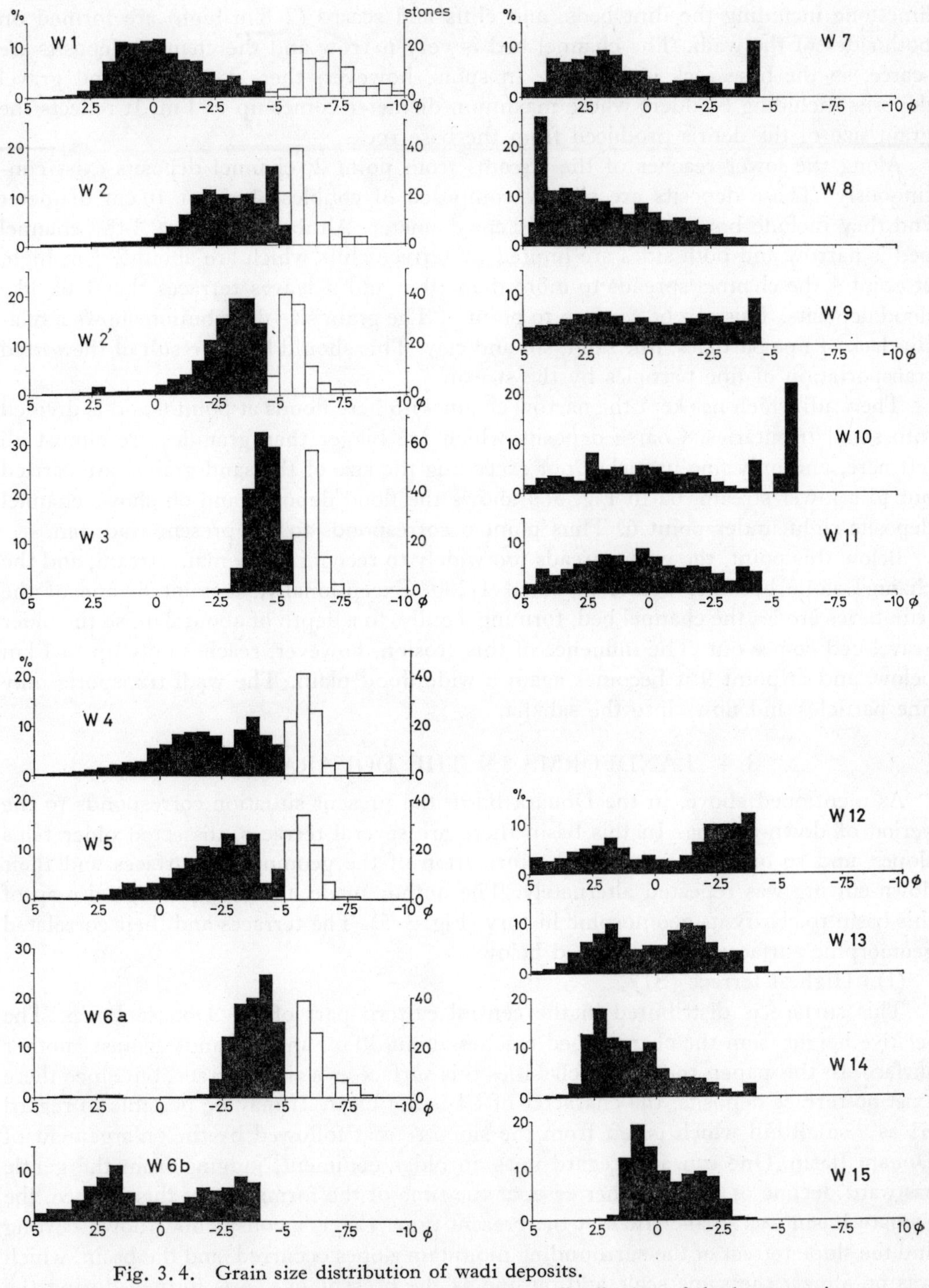

Fig. 3. 4. Grain size distribution of wadi deposits.
White columns show the frequency distribution of the longest axis of the biggest one hundred gravels per square meter. Black columns shows the weight composition of rest fine materials.

limestone including the flint beds, and cliffs and scarps (2–8 m high) are formed on both sides of the wadi. The channel bed is very narrow and the channel deposits are scarce, so the base rock is exposed. In spots, however, there exist sand and gravel deposits including boulders whose maximum diameter comes up to 1 m. It reflects the grain size of the debris produced from the base rocks.

Along the lower reaches of the streams from point 2, channel deposits exist continuously. These deposits are mainly composed of cobbles of about 10 cm diameter and they include boulders of about 50 cm diameter. Although at point 2 the channel bed is narrow and both sides are limited by terrace cliffs which are about 2–5 m high, at point 4 the channel spreads to more than 10 m and it leaves terraces that look like flood deposits. This can be seen up to point 6. The grain size distribution shows a relative lack of fine particles, fine sand, silt and clay. This should be the result of the sorted transportation of fine particles by the stream.

The wadi which has kept the narrow channel up here floods at point 6 and is divided into small tributaries. Coarse deposits which are bigger than granules are almost all left here, and only fine particles, not exceeding the size of the sand grains, are carried out to a lower stream. 6a in Fig. 3. 4 shows the flood deposits and 6b shows channel deposits right under point 6. Thus point 6 corresponds to the present wadi fan.

Below this point, the wadi spreads too widely to recognize the main stream, and the channel slope becomes gentle to about 1/200. Exceptionally, at point 8, one of the tributaries erodes the channel bed, forming a gully, to a depth of about 2 m, so the older gravel bed comes out. The influence of this erosion, however, reaches only up to 1 km below, and at point 9 it becomes again a wide flood plain. The wadi transports only fine particles and flows into the sabkha.

3. 4 LANDFORMS IN THE DOUARA BASIN

As mentioned above, in the Douara Basin the present situation corresponds to the period of down-cutting. In this basin there are several terraces, dissected older talus slopes, and so on. It seems that the formation of the geomorphic surfaces and their down-cutting was repeated alternately. The author made a geomorphological map of this basin to clarify its geomorphic history (Fig. 3. 5). The terraces and their correlated geomorphic surfaces are mentioned below.

(1) Highest terrace (S1)

This surface is distributed in the central eastern part of the Douara Basin. The relative height from the channel bed reaches about 30 m, and it stands against another surface. In this paper, the author classifies this surface as a river terrace, but since there exist no terrace deposits, the character of S1 is not clear. It may be possible to regard S1 as a small hill which is left from the slope retreat followed by the enlargement of Douara Basin. One can also regard it as an older pediment, judging from the gentle eastward decline of S1. In either case, at the time of the formation of this surface, the Douara Basin was smaller than at the present time. Later, a conspicuous down-cutting and the slope retreat of the surrounding mountain slopes occurred, and the basin, which was on almost the same scale and outline as the present one, was formed during the period of higher terrace (S2) formation.

On the S1 there exists a weathered horizon and a pinky, light-brown soil horizon

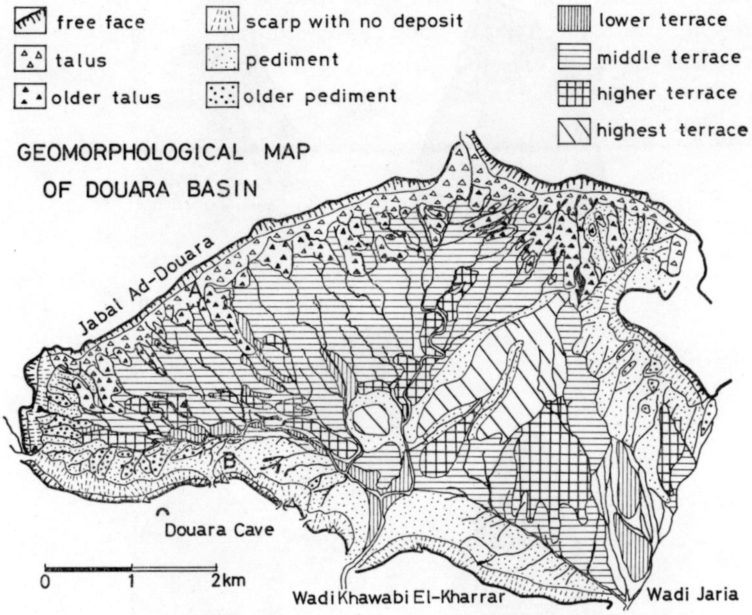

Fig. 3. 5. Geomorphological map of the Douara Basin.

composed of well-sorted sand and silt, both of which are 15 cm in thickness. A lot of gravel and flint artifacts and their fragments are scattered on this surface. This surface had been dissected by the wadis, and the terrace cliff, caused to retreat by erosion, changes into a concave pediment.

(2) Higher terrace (S2) and its comparable geomorphic surfaces

The higher terrace is distributed along the wadi at the central part of the Douara Basin. It is about 10–12 m above the channel bed, and it gradually increases in height toward the mountain side. The edge of S2 is rounded off by erosion. It is an accumulation terrace with a deposit 2–3 m thick. The deposits consist of sand and gravel particles which are mainly of middle size, but sometimes as large as 10–20 cm. On the S2 terrace and gentle terrace slope many flint artifacts and their fragments are found. They include handaxes of the Acheulian type of the Lower Paleolithic (Plate 3. 2) and artifacts of Levalloiso-Mousterian Middle Paleolithic type. And also in the terrace deposits handaxes are found.

Declined geomorphic surfaces, shaped like monadnock, are scattered at the foot of the cliff which limits the Douara Basin in the north (Plate 3. 3). The surfaces consist of angular gravel 5 m thick, which sits on the base rock with disconforming. The deposits include silt, sand, and gravel which sometimes reaches 1–2 m in size, and there are a lot of gaps within the gravel. The lower portion of this deposit is hardened, and indistinct layers are observed. The angular gravel in the lower portion is so weathered that it is frail. According to this observation, the surface is assumed to be of the talus which was deposited at the stage of regression of the cliff, prior to the present location. The surface of the talus is about 15 m above the channel bed at the cliff side, and gradually reduces its height toward the central part of the basin, and finally this surface

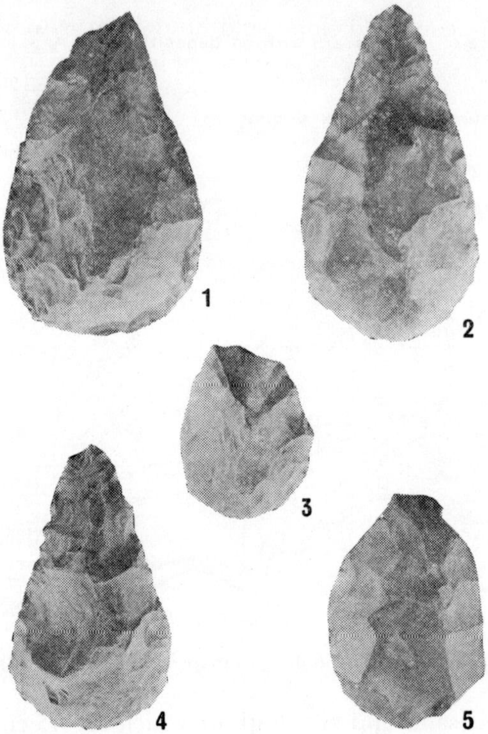

Plate 3. 2. Acheulian handaxes discovered at the point 60. ×1/4.

Plate 3. 3. Older dissected talus slope.

is cut off by the middle terrace surface (S3), which is described later. Considering these characteristics of landform and deposits, the surface surely seems to be correlated to S2.

There is a continuous cliff at the southernmost side of the basin. This cliff forms the north facing slope of the mountain ridge which shows an asymmetric form, and the Douara Cave is located on the southern slope of the ridge. The lower part under the cliff forms a rather gentle concaved slope, and no deposit is found on the slope. There is a terrace-like geomorphic surface which protrudes 6–7 m from the gentle slope like a monadnock. There are thin deposits there, consisting of eolian fine sand and silt 20 cm thick, but no gravel is seen. This geomorphic surface is obviously the remnant from the regression of the slope, and it is the rest of the former pediment.

On the surface of this geomorphic surface, many artifacts of Levalloiso Mousterian and their fragments are scattered about several tens of square meters. They are assumed to be the remains of a site where people used to make tools. Near the remains, outcrops of flint beds are observed, and they are considered to have been used as the materials for the tools. In addition, a few handaxes of the Acheulian type are collected at one point.

Considering the height from the channel bed, this surface seems to be correlated to S2.

(3) Middle terrace (S3) and its correlated surface

The middle terrace surface, which is about 3–4 m above the channel bed, is the most widely distributed terrace in the Douara Basin. This terrace is an accumulation terrace with deposits of sand and gravel 2–4 m thick. Prior to the accumulation of the terrace deposits, S2 had been graded, about 6 m downward and considerably to the side, to form a wide valley. As a result of this grading, the bed of the Douara Basin was considerably lowered, and the distribution of S2 became sporadic. As a cause of the downcutting of the base surface in the basin, it is possible to assume that a steram piracy by the Wadi Khawabi el Kharrar which wrecked the mountain ridges on the south boundary of the basin, has occurred against the upper reaches of the Wadi Jaria which flowed eastward in the Douara Basin at that time.

The deposits of S3 include small gravel, which is usually 1–3 cm in diameter, but sometimes as large as 5–20 cm in diameter, and rarely contain matrix. Stratification is obscure and the lower part of the deposits is slightly hardened. The surface of the deposits has become reddish-brown in color because of weathering. The terrace deposits of S3 include many Levalloiso Mousterian artifacts and their fragments. However, they are not found on the surface, and there is no indication that people used to make tools here. Although some flint artifacts are found on the surface at the points 35 and 59 in Fig. 3. 11, they are rather new and they belong to the Pre-Pottery Neolithic Period.

Levalloiso-Mousterian flint artifacts are distributed on S3 only in the area around S2 and sporadically along the west side of the basin. These surfaces seem to be the places of early emergence from water among S3 terrace groups. On these surfaces, the deposits are rather thin, being 0.5–1.5 m thick, and are reddish-brown in color due to weathering. The lower part of the deposits is hardened. They mainly consist of sub-angular gravel particles 2–7 cm in diameter, with sandy matrix. They are assumed to be a strath terrace formed at the time of the erosion of S2.

S3 continues smoothly to the talus slope under the cliff on the northern side of the basin, and also to the concaved slope under the cliff on the southern side. These two slopes are considered to have been formed at the same time as S3. The surfaces of these slopes are cut down continuously by the wadi flow, likewise at S3. Gullies are formed on the surfaces, but the dissections are not so advanced, and the surfaces keep their original state rather well. The remains of a flint factory site are not found on these surfaces. This fact is important in considering the age of the formation of these slopes.

There is a terraced wadi fan on the exit of the Douara Basin formed by the Wadi Khawabi el Kharrar. It also seems to have been formed in the same period as S3.

(4) Lower terrace (S4)

The lower terrace is the low accumulation terrace 0.5–1.5 m above the channel bed. It is distributed along the main course of the wadi and along rather large tributaries, but not along the upper reaches of the wadi which is narrow and of the gully type. The terrace deposits are 0.5–1.2 m thick. Although the terrace can be said to be a kind of accumulation terrace, it can be called a strath terrace in some places where the wadi cuts the base rock. In any case, the sedimentation period is rather short. The deposits are finer than those of S3 and consist of the alternation of the strata including mainly gravel particles 3–6 cm in diameter and of sandy strata with gravel. The gravel is rather fresh and no concretion of the deposit is observed.

Detailed observation of S4 shows that 2–3 terraces can be recognized. The remains of the flint factory site are found only on the uppermost surface of S4. The other surfaces are on new deposits and the lowermost surface seems to have chances of getting water during the annual flooding period. No flint artifacts are found on these lower surfaces.

3. 5 ENVIRONMENTAL CHANGES IN GEOMORPHIC SURFACE FORMATION

At least two formative periods in the vast accumulation terrace are recognized in the Douara Basin, and there are down-cutting periods before and after the accumulation periods. The environmental changes in geomorphic surface formation will be described as an example below of those from the formative period of S3 to the down-cutting period.

(1) Formative period of S3

S3 is an accumulation terrace, with a few exceptions. The eroded detritus from the mountain slopes and S2 terrace buried the vast valley to become the accumulation terrace. In this period it is considered that a great amount of debris was produced, and the vast valley gradually changed to a flood plain. The debris was mainly supplied from the cliff that surrounds the Douara Basin. That is, the accumulation period is assumed to coincide with the regression period of the cliff. The mode of regression of the cliff seems to have been a parallel retreat, and supply of the debris is considered to have occurred mainly by crumbling.

With the progress of regression of the cliff, thick taluses were formed under the cliff on the north side of the basin, and concave slopes having few deposits were formed on the south mountain side. Both of these types of slopes continue to S3, and are considered to be contemporaneous. Since the field data are not sufficient to discuss the characteristics of the cliff, the formation of these two slopes will mainly be dealt with.

Talus and pediment

As examples of taluses and concaved slopes, the vertical profiles of the slopes at points A and B on the Fig. 3. 5 are shown in Fig. 3. 6. At point A, the talus formed below the cliff which is 70° to almost 90° in gradient, and the talus slope has an almost straight or slightly concave surface which dips 28°–32°. The cross-section of the talus

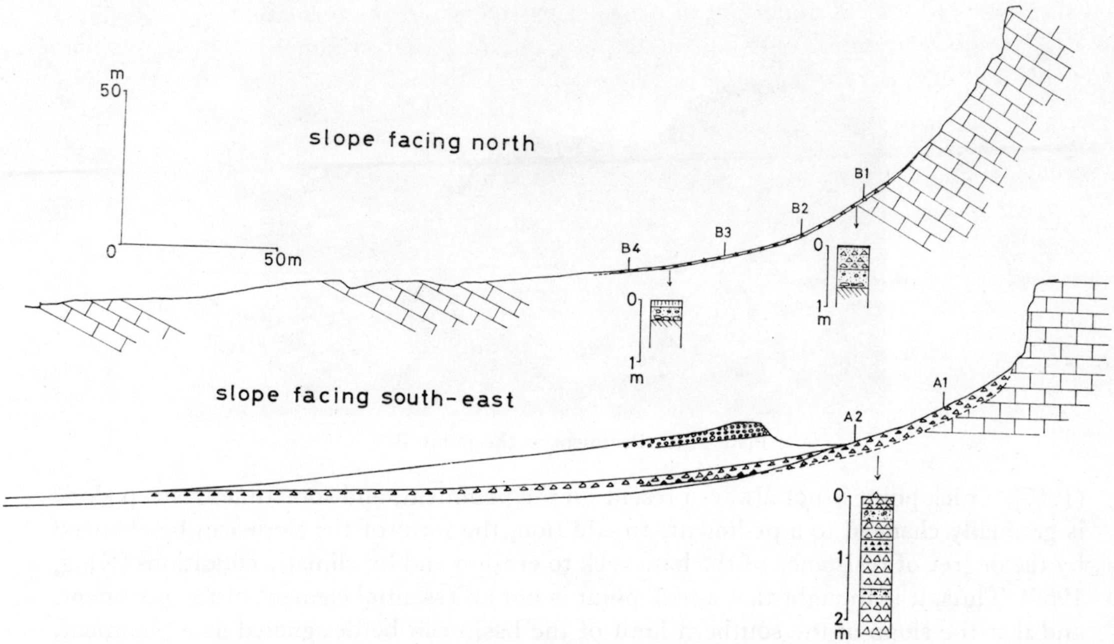

Fig. 3. 6. Vertical profiles of two piedmont slopes.

slope is about 150 m and its lower part continues to S3. The deposits consist of an alternation of strata including mainly angular gravels of 20–30 cm, sometimes 2 m in diameter, and of fine sand-silt strata with angular gravels 2–3 cm in diameter. The depth of the deposits is estimated to be more than 5 m.

At point B, however, the slope has a smooth surface with a much more concave curve. The upper part of the slope is 32° in gradient, the middle part is 17°–18°, and lower part is from almost even to 6°. The cross-section of the slope shows a concave curve, and the contours of the slope appear to be concentric circles. The thickest deposit is only 40 cm, and the matrix mainly consists of light brown sand with angular gravel a few cm in diameter. Upon this layer there are large, rough boulders fallen down from the cliff, and at first glance it appears to be a debris slope. These boulders fall down to a rather lower part of the slope, but do not reach the gentle slope, which is less than 6° in gradient. The boulders extend about 200 m downward from the base of the cliff. The size of the boulders is gradually reduced toward the gentle slope, in a reverse order tendency compared with the case of the talus. The largest one is 2 m in diameter (Fig. 3. 7). The base rocks are exposed on the higher parts of the slope, and are sometimes exposed on the lower parts. This slope is obviously an erosion surface cutting the base rock. Considering the landforms and characteristics of the deposits, the slope can be designated as a pediment.

According to the literature, however, this type of slope is usually made up of two parts separated by a rather distinct nick point, while the upper part is a steep slope of 20°–40° in gradient, and the lower part is the pediment of a gentle slope (Akagi, 1972). In the Douara Basin, however, the nick point is rather indistinct. According to Twidale

Plate 3.4. Pediment at the point B.

(1967), a nick point is not always present on the pediment, and sometimes a steep slope is gradually changed to a pediment. In addition, the form of the slope can be changed by the degree of resistance of the base rock to erosion and by climatic conditions (King, 1968). Thus, it is thought that a nick point is not an essential element of the pediment, and that the slope of the southern limit of the basin can be designated as a pediment. In the Douara Basin, the slope is divided into an upper slope covered with debris and a smooth lower slope. The upper slope is comparable to the backward steep slope mentioned by Akagi (1972) and to the debris slope mentioned by King (1953). The lower slope is a pediment in a narrow sense. On the geomorphological map of this report (Fig. 3. 5), the slope of point B is divided into the upper steep slope and the pediment. In describing the slope of point B, however, the term "pediment" is used in a broad sense, including both upper and lower surfaces.

Talus formation requires a great volume of debris supply from upside cliffs surpassing debris removal by the stream on and at the foot of the slopes. Thus, the conditions enabling the production of such debris will be discussed and compared with present condition.

The cliff which produces debris is steep and appears to be a wall. At a glance, it seems to be on the point of breaking up, but the surface is rough and slaggy, showing etching by corrosion. Hardly any traces of young breakage on the surface can be observed. Though such weathered forms are characteristic of limestone, it is considered that chemical weathering has been dominant on the surface of the base rock for a long time and that mechanical weathering and debris production due to reasons such as frost action and saline weathering (Ollier, 1969) are indistinct. In addition, young debris from the cliff is seldom recognized on the talus slope or pediment where the debris is thought to remain. On all of the cliffs around the Douara Basin, the deposition of young debris is observed only at three places. The author did not come upon a breakage of the cliff or a rock fall during his investigation. As he did not study the area in winter, it is impossible to conclude that these do not occur. Judging from the above-mentioned facts, however, debris production is considered to be indistinct under present climatic conditions. Moreover, the debris on the slopes changed in color, and the surface is overlaid

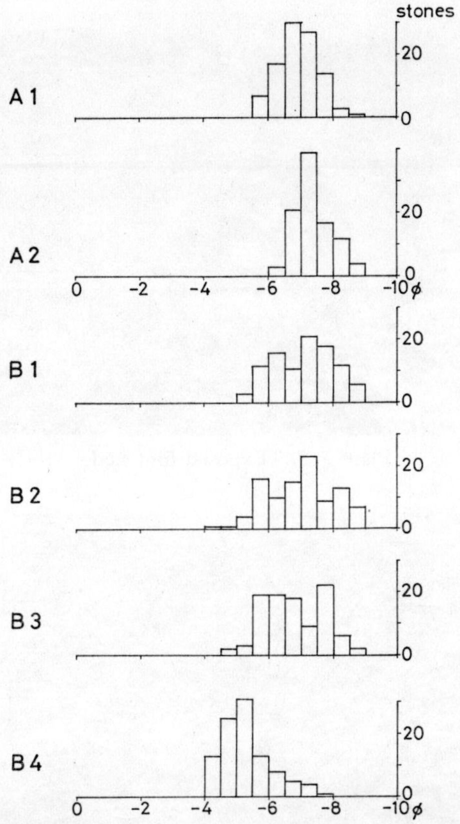

Fig. 3. 7. Frequency distribution of the longest axis of gravels which cover the surfaces of talus slope and pediment (the biggest one hundred pieces per square meter).

with soil. Thus, the present situation can be regarded as being stable (Fölster, 1969). It can be considered, in addition, that the growth of talus has nearly stopped.

This consideration is also indicated by the talus deposits. A lot of snail shells were found from the II horizon and IV horizon composed of fine materials in the debris profile at point A, as shown in Fig. 3. 8. Carbon 14 dates of each snail shell horizon show 4180±100 y.B.P. (N-2254) and 11300±160 y.B.P. (N-2255). According to Rubin, Likens and Berry (1963), when snail shells are used, the carbon 14 dating is sometimes older than the true age and the date is not always reliable.[4] Therefore, these dates can be used only as references. It is certain, however, that the supply of young debris is small and that most of the talus debris might have been supplied before the Holocene Age. The accumulation of snail shells means that the slope has been stable for a long time. During that time only very fine materials have been supplied in small quantities and the slope has been covered with soil.

Judging from the above, it can be concluded that the good supply of debris forming

4. It is possible that snail shells were intermixed with carbon 14 originating from the base rock. In this case carbon 14 dating is said to be older than the true date by 1000 years.

Plate 3. 5. Exposed flint bed.

Plate 3. 6. Flint strewn field.

the talus resulted from different climatic conditions from those at present. Those conditions seem to have been colder and rainier,[5] as stated by Wirth (1958) and Abdul-Salam (1965). The good supply of debris may have based on the intensive frost action under such conditions.

In the Palmyra region there is also frost action in winter at present, as mentioned in the second section. The precipitation in winter is considered to give favorable conditions for the destruction of rocks and the production of debris. Nevertheless, the rather small production of debris may be due to the small amount of frost action and to the feeble intensity of freezing.[6]

It is reported that production of debris has increased around the Lebanese High Mountains, and that talus and other depositional landforms were formed in various

5. Evidence of rainy condition will be shown when the deposits of the wadi are mentioned.

6. As there are few reports about the relation between the intensity of freezing and debris production, it cannot be understood in detail. But it is certain that intensive freezing, involving low temperatures and much moisture, is required in debris production.

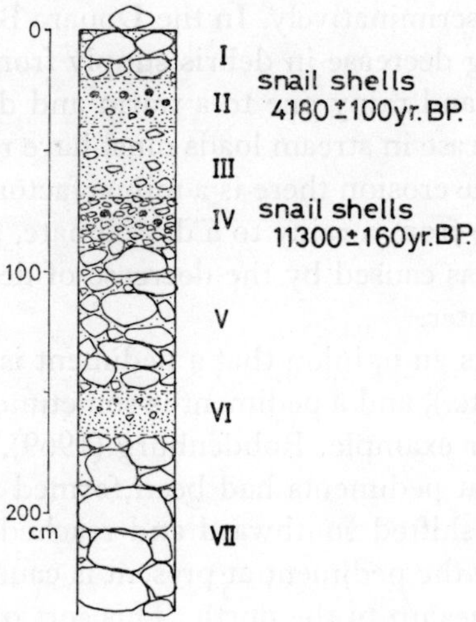

Fig. 3. 8. Vertical section of talus deposit at the point A.

parts of these mountains (Kaiser, 1963; de Vaumas, 1970; Besançon, Dresch et Tricart, 1973; etc.). Moreover, the lower limit of the periglacial zone is supposed to go down to 850 m above sea level in the Palmyra region (Kaiser, 1969; Farrand, 1971). For these reasons, if Douara Basin did not belong to the periglacial area, it is considered that it had similar conditions to those of the periglacial zone.

During the same age the formation of the pediment also progressed. The supply of debris from the cliff must have increased on the pediment as well as on the talus slope. Nevertheless, a talus was not formed, but a pediment was. One of the reasons may be the relatively small supply of debris on the pediment compared with the talus slope, and the other may be a strong sheet wash caused by increased rainfall. Under these conditions, the removal of debris on the slopes and denudation must have progressed. A pediment, so to speak, was formed under the equilibrium between the debris supply and the removal. If the condition had varied slightly, the pediment would have changed to a talus or it would have been dissected by linear erosion.

The reason why the quantities of the debris supply were different in each slope below almost the same cliff is not clear. One of the causes considered is the difference in the resistance between two basal rocks in each place. The horizon at point B, which lies in stratigraphy higher than that at point A, may be more resistant. The other is the difference in the decline of a base rock in each place. Compared with the almost horizontal stratum at point A, the stratum declines southwards at point B. Another cause considered is the difference in the direction of each slope, followed by the difference in geomorphic processes. They may have caused the difference in the quantities of debris supply.

At present, both talus slopes and pediments are dissected from the end of the slopes by a gully erosion and a part of them becomes a terrace. Thus the present age corresponds to the descriptive phase of these surfaces. As the cause of this down-cutting, Twidale (1967) asserted that the process is as follows: a remarkable weathered zone was formed at the foot of the slope by the penetration of water flowing down from the upper slope and it was eroded discriminatively. In the Douara Basin, however, the cause considered at first is the striking decrease in debris supply from the cliff, caused by the fchange o climate from a cold and rainy one to a warm and dry one, accompanied by the end of the glacial age. Decrease in stream loads must have revived the erosion and led to down-cutting. To revive the erosion there is a minus factor such as a decrease of flowing water following the change from a rainy to a dry climate, but it is explained that the progressive down-cutting was caused by the decrease of debris supply surpassing the decrease of the volume of water.

Moreover, there is an opinion that a pediment is likely to be created in an arid land (Mensching, 1958, etc.), and a pediment is sometimes used as a facies landform in much of the literature. For example, Rohdenburg (1969), who studied the pediments in Nigeria, considered that pediments had been formed during the Würm glacial age when the Saharan Desert shifted southward and reached Nigeria. His interpretation is that the linear erosion on the pediment at present is caused by the increase of rainfall owing to NITC which comes up to the north. This sort of interpretation is adopted by many scholars, such as Bigarella and Moushinho (1966) and Burke and Durotoye (1971). In the Douara Basin, however, in spite of the fact that it is located in an arid land, the pediment is in a destructive phase at present, and pedimentation proceeds more in rainy periods. The case in the Douara Basin may be peculiar, but it might not be advisable to connect the pedimentation directly to an arid climate.

Transportation by Wadi Khawabi el Kharrar during the Würm Glacial Age

In former section, the transportation by Wadi Khawabi el Kharrar at present was described. Although all the coarse materials which are bigger than pebbles are deposited soon after the wadi goes out of the Douara Basin, it could be said that those coarse materials had been transported to the sabkha during the age of the supply of the gravel which would have formed the S3 deposits. That gives very interesting information for guessing not only the environment but also the geomorphic processes and the climatic conditions in the past.

At point 14 where the wadi is about to flow into the sabkha, consolidated round or subangular shaped pebble (diameter: 2–3 cm) beds are observed under the present wadi deposit, 30 cm in thickness, composed of fine sand and silt. At points 8 and 10, weathered light brown or yellowish-brown pebble (diameter: 4–8 cm) beds are also observed under light brown or yellowish-grey sand and silt beds which are from 30 cm to 80 cm in thickness. There is a surface of small unconformity between these two horizons. While there is no reason to correlate the deposit at point 14 with the S3 terrace deposits, the gravel beds at points 8 and 10 are definitely correlated with S3 deposits, based on the following facts.

At point 6, the wadi fan (F3), which is built at the outlet of the Douara Basin and continued from S3, intersects with the present wadi fan which corresponds to an alluvial surface. The farther it is from the outlet of the Douara Basin, the lower the height of F3 becomes, and at point 6, F3 enters beneath the present wadi fan. Thus, the slope gradient of that time was more abrupt than that of the present. The weathered gravel beds at both point 8 and point 10 are buried on an extensional part of F3 deposits, and the sand and silt horizon overlying them is the present wadi fan deposits. Therefore, at the time when gravel composing F3 deposits was sedimented, it is certain that gravel

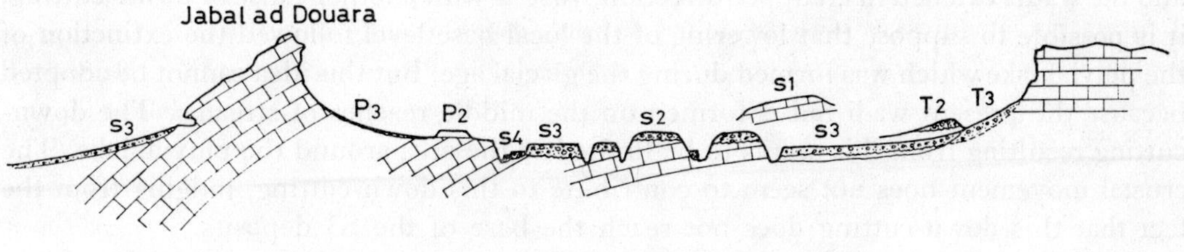

Fig. 3. 9. Schematic profile of the Douara Basin.

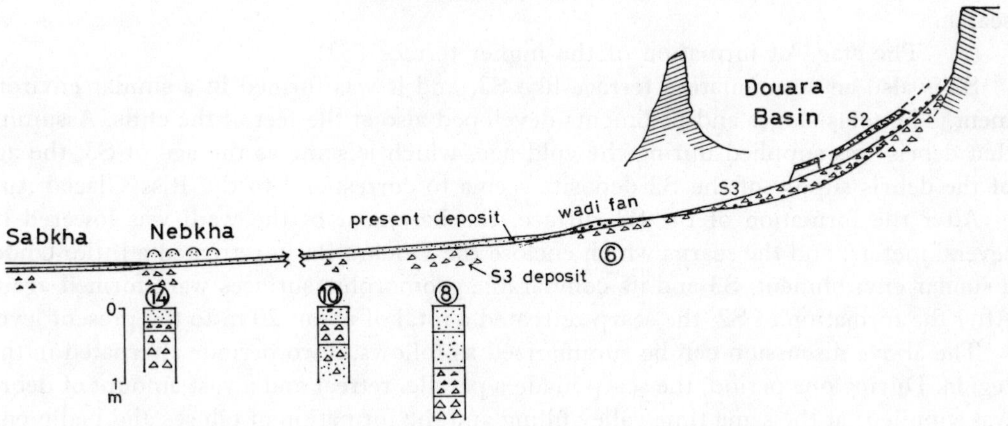

Fig. 3. 10. Geomorphic process and schematic profile along the Wadi Khawabi el Kharrar.

had been transported to the lower reaches of the streams beyond the limit of the present gravel distribution. The gravel bed at point 14 may be correlated with that of point 8. In Fig. 3. 10 the consideration mentioned above is shown schematically.

It may have been a rainy climate that made it possible to transport the debris to the sabkha. Although this region belongs to a semi-desert now, it is considered that it was formerly far more rainy than present, and that a great deal of water had run off. It is impossible to assume a seasonal variation of rainfall, but it can be assumed that there were perennial streams. Because a number of flint artifacts are included in the S3 terrace deposits, it is estimated that many people lived in this region at that time.

In spite of the active fluvial process, a vast accumulation arose in the upper reaches of the stream. This fact shows that debris production on the mountain slopes was extremely active. The density of vegetation is estimated to have increased considerably owing to the increase in rainfall, but it is impossible to examine this phenomenon because of the lack of data, such as data from pollen analysis. The role of vegetation in connection with the slope retreat, however, seemed not to be very important.

(2) Terracing of the middle terrace (S3) — Down-cutting stage —

The cause of down-cutting of the S3 flood plain, which continued to accumulate deposits for a long time, may be the striking decrease in the debris supply, similar to the down-cutting of the talus slope and pediment mentioned above. This down-cutting is estimated to have begun at the end of the glacial age. Although after that there were two or three stagnation phases, the basic tendency of down-cutting did not alternate and the wadis extened in an upper direction. Just as with another cause of down-cutting, it is possible to suppose that lowering of the local base level followed the extinction of the pluvial lake which was formed during the glacial age. But this idea cannot be adopted because the present wadi fan is formed on the middle reaches of streams. The down-cutting resulting from this seems to be limited to the area around the pluvial lake. The crustal movement does not seem to contribute to this down-cutting, judging from the fact that this down-cutting does not reach the base of the S3 deposits.

Unfortunately, the causes and the environment of the formation of the lower terraces cannot be clarified, owing to the season in which the author carried out this research.

(3) The stage of formation of the higher terrace (S2)

S2 is also an accumulation terrace like S3, and it was formed in a similar environment. The talus slopes and pediments developed also at the feet of the cliffs. Assuming that debris was supplied during the cold age, which is same as the age of S3, the age of the debris supply of the S2 deposits seems to correspond to the Riss Glacial Age.

After the formation of the S2 surface, the basement of the wadi was lowered by several meters, and the scarps which enclose the Douara Basin retreated a little. Under a similar environment, S3 and its comparable geomorphic surfaces were formed again. After the formation of S2, the scarp retreated a total of about 20 m to the present level.

The above discussion can be summarized as follows. Two periods alternated in this region. During one period, the scarp made a parallel retreat and a vast amount of debris was supplied; at the same time valley filling and the formation of taluses and pediments continued. During the other period, linear erosion was dominant, and the terracing and the dissection of the talus slope and pediment took place. Those periods seem to correspond to a cold, rainy glacial age and a hot, dry interglacial age, respectively. Along the wadi, a wadi fan which was wider and steeper than that at present was formed, owing to the increase in debris supply and to the strong power of the current during the glacial age. On the other hand, during the interglacial age, the supply of debris decreased and only fine materials, smaller than sand grains, were transported to the sabkha and sedimented there. Eolian landforms include nebkha, and sand dunes are said to be the landforms of the interglacial age.

3. 6 AN ESTIMATION OF THE AGE OF EACH TERRACE FROM THE FLINT ARTIFACTS

In the pervious section, the history of the natural environment was described through the geomorphic development. However, the age of each surface was only speculative. Fortunately, in this area many kinds of flint artifacts from the Paleolithic age to the Neolithic age have been discovered, so one can trace its transition almost continuously. It is possible to trace each surface by using many of the flint artifacts. Like volcanic ashes, flint artifacts do not indicate directly their absolute ages, but they are very effective in determining the relative ages of some geomorphic surfaces and deposits if they are on each surface or if they are included in some deposits.

First of all, the places where the flint artifacts were discovered are shown in Fig. 3. 11. Some places in Fig. 3. 11 are considered to have been the flint factory sites. Though the distribution of the Lower Paleolithic artifacts, mainly composed of Acrelian hand-

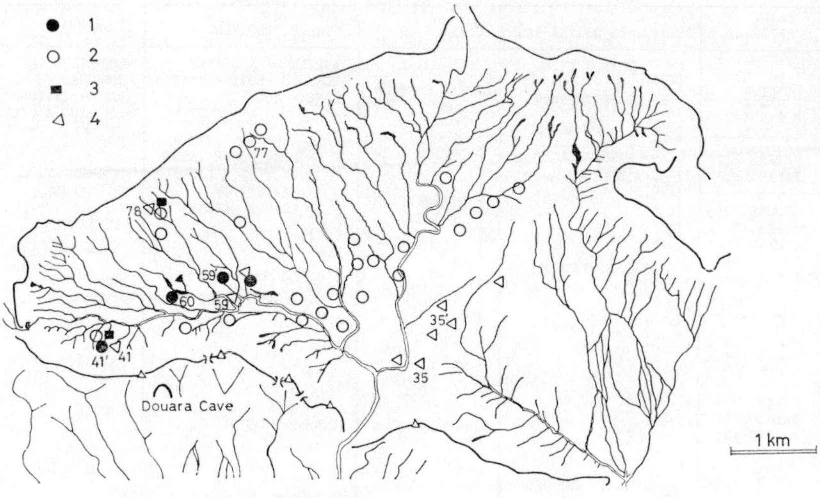

Fig. 3. 11. Distribution of flint artifacts and flint strewn fields.
1. Lower Paleolithic, 2. Middle Paleolithic, 3. Upper Paleolithic,
4. Neolithic.

axes, are restricted to several points, 59′, 60, 41′ etc., Middle Paleolithic artifacts, mainly composed of Levalloiso Mousterian types, are distributed widely around the outcrops of the flint beds. The only Upper Paleolithic site discovered is at point 78. Pre-Pottery Neolithic tools (PPNA, PPNB) were found at several points, 33, 35, 35′ etc., near the confluence of two branches of the wadi and at the points 59, 59′, 77, 78. Microflint tools and rubbed flint tools which belong to Neolithic artifacts have not yet been found, however.

We shall now investigate the relationship between the distribution of flint tools and that of the geomorphic surfaces. Comparing Fig. 3. 11 with Fig. 3. 5 (Geomorphological map of the Douara Basin), it becomes clear that the distribution of the Acheulian handaxe is restricted to the S2 and its comparable surfaces. The handaxe is not found on surfaces lower than S2. From this fact, it is presumed that the terracing of S2 had already been finished before the end of the Acheulian culture, which belonged to the last stage of the Lower Paleolithic age. The age of accumulation of S2 terrace gravels is also included in the age of Acheulian culture, for handaxes are discovered in the terrace deposit of S2. On the S2 surface many kinds of flint artifacts which belong to the period from Levalloiso Mousterian to PPNA or PPNB are found also. In the previous section the author surmised that the age of the S2 deposit corresponds to the Riss Glacial age. This presumption may be adequately supported by the flint artifacts.

The distribution of the Levalloiso Mousterian artifacts is also almost restricted to S2. They are not found on the S3 except in special areas. The flint artifacts that were observed belonged to PPNA or PPNB without exception. This indicates that S3 was a wide flood plain during the Levalloiso Mousterian cultural age and that the terracing of S3 began after the end of this culture. The fact that S3 deposits include Levalloiso Mousterian flint fragments supports this observation. Levalloiso Mousterian flint artifacts are not observed on the pediment and talus slope which are comparable to S3.

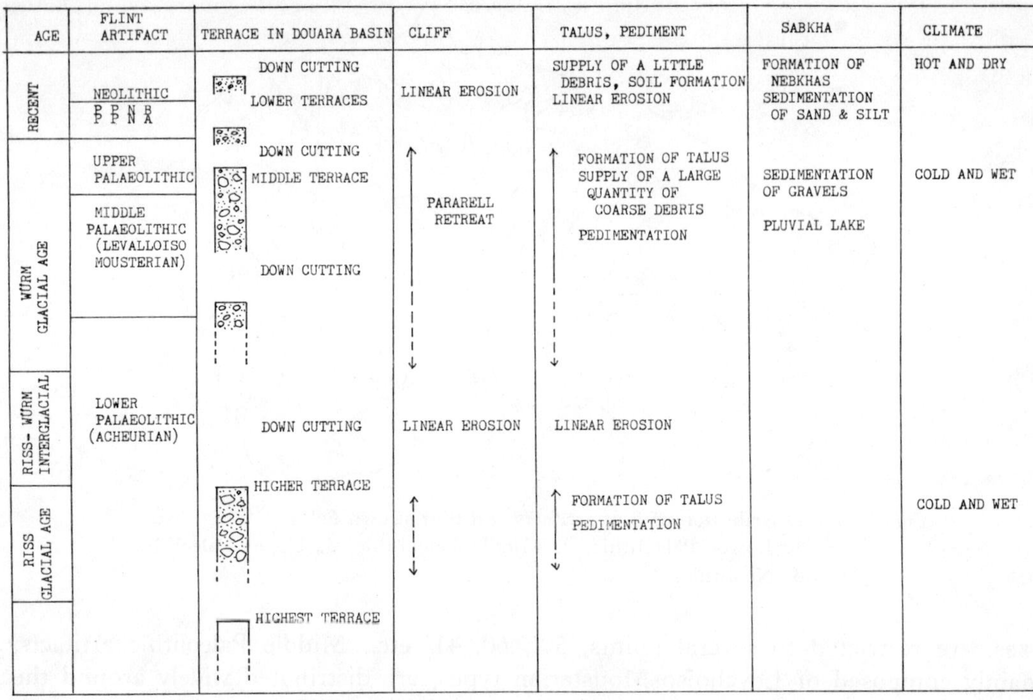

Fig. 3. 12. Chronological arrangement of terraces, talus slopes and pediments.

In the same way as S3, the formation of these slopes seems to have continued after the end of the Levalloiso Mousterian culture. However, on the piedmont part or the marginal parts of S2 which are comparable to S3, a number of Levalloiso Mousterian flint artifacts have been discovered. This indicates that those parts had become terraces prior to the end of the Levalloiso Mousterian cultural age.

In general, there are no flint artifacts on the lower terraces (S4), but flint tools belonging to PPNA or PPNB were discovered at several places such as point 35. These places are located on the uppermost terrace of the lower terrace group. This fact indicates that this surface became a terrace before the PPNA cultural age. The terracing of the other lower terraces is considered to have taken place after the PPNA cultural age, but the details are not clear.

From the above-mentioned facts, the age of accumulation of the S3 gravel and the beginning of the down-cutting of S3 can be clarified. Considering the climatic conditions again, the age of the accumulation of the S3 gravel seems to coincide with the Würm Glacial age, and the terracing of S3 must have begun tens of thousands years ago in the late glacial age.

The chronology of this region is shown in Fig. 3. 12 according to the above-mentioned considerations.

3. 7 SUMMARY

1. In this paper the author dealt with the history of the natural environment in the late Quaternary in and around the Palmyra region which is located in the northern part

of the Syrian Desert. Mainly studied is the area between Jabal ad Douara, which is the mountain range bounded on the south by the Palmyra Basin, and the playa lake in the central part of the Palmyra Basin.

2. There is a small basin named the Douara Basin in Jabal ad Douara. Along the wadi arisen in this basin, the author picked out the wadi deposits and analysed the size distribution of these deposits in order to examine the transportation of wadi deposits under the present semi-desert climatic conditions. Soon after the wadi leaves the Douara Basin, it floods and deposits the coarse materials which are bigger than granules to make a wadi fan. It is mainly fine sand or silt which are transported by the wadi down to the sabkha (playa). In the marginal zone of the sabkha, countless numbers of dune-like small mounds, called nebkha, which are composed of fine sand, silt, and salt grains blown from the sabkha, are seen.

3. In the Douara Basin, two marked accumulation phases are recognized. The deposits are composed of the debris derived from the mountain slopes. These two deposits are dissected to form terraces, with a higher terrace and a middle terrace at present. There are talus slopes and pediments as the comparable geomorphic surfaces of the two terraces. Both the talus slopes and the pediments are being dissected at present. During the age of debris supply for the accumulation terraces, the wadi fan, which is dissected at present, was formed at the outlet of the Douara Basin. The gravel bed which composes this wadi fan enters beneath the deposit of the present wadi fan and this gravel bed is traced to the sabkha. This indicates the possibility of gravel transportation to the sabkha during the age of the formation of the older wadi fan. The author considers that a cold rainy climate makes possible the supply of a large amount of debris and the transportation of pebbles to the sabkha. Such climatic conditions possibly imply a glacial age, and the two accumulation periods are presumed to correspond to the Würm and Riss Glacial ages.

4. In addition to the landform survey, the relationships between the distribution of flint artifacts and each geomorphic surface were examined. The distribution of Acheulian handaxes is limited to the higher terrace and its terrace deposit. This indicates that the terracing of the higher terrace had progressed before the end of the Acheulian cultural stage. On the other hand, there are no Levalloiso Mousterian flint artifacts on the middle terrace, with a few exceptions, and they were only discovered from the terrace deposit. Thus, the terracing of the middle terrace can be thought to have begun after the end of Levalloiso Mousterian cultural age. Pre-Pottery Neolithic artifacts are observed on the uppermost surface of the lower terrace group and on the surfaces higher than that. This indicates that the down-cutting of the middle terrace had already begun prior to that time. These results are not inconsistent with the age of each geomorphic surface as surmised from the geomorphological survey.

LITERATURE CITED

Abdul-Salam, A. (1965) Morphologische Studien in der Syrischen Wüste und dem Antilibanon. Berliner Geogr. Abh. H. 3.

Akagi, Y. (1972) Pediment Morphology in Japan. Bull. Fukuoka Kyoiku Univ., **21**: Pt. 2. 1–63.

Besançon, J., Dresch, J. et Tricart, J. (1973) Observations sur les processus morphogenetiques <froids> au Liban. Rev. Geogr. Phys. Geol. Dynamiq., (2). **XV**: 231–272.

Bigarella, J. and Mousinho, M. (1966) Slope development in Southeastern and Southern Brazil. Zeitschr. f. Geomorph., **10**: 150–160.

Burke, K. and Durotoye, B. (1971) Late Quaternary climatic variation in Southwestern Nigeria: evidence from pediments and pediment deposits. INQUA VIII Congrés (Paris 1969) Etud. Quatern. Monde., 145–153.

Cooke, R. V. and Warren, A. (1973) Geomorphology in Desert. 317 pp. Batsford Ltd., London.

Denny, C. S. (1967) Fans and Pediments. Am. J. Sci., **265**: 81–105.

Farrand, W. R. (1971) Late Quaternary Palaeoclimates of the Eastern Mediterranean Area. The late Cenozoic Glacial Ages. Yale Univ. Press, 529–564.

Fölster, H. (1969) Slope development in SW-Nigeria during Late Pleistocene and Holocene. Gies. Geogr. Schr. H. 20. 3–56.

Glennie, K. W. (1970) Desert sedimentary environments. Development in Sedimentology. 14. 222 pp.

Holm, D. A. (1960) Desert Geomorphology in the Arabian Peninsula. Science, **132**: 1369–1379.

Kaiser, K. (1963) Die Ausdehnung der Vergletscherungen und "periglazialen" Erscheinungen während der Kaltzeiten des Quärtären Eiszeialters innerhalb der Syrisch-Libanischen Gebirge und die Lage der klimatischen Schneegrenze zur Würmeiszeit im östlichen Mittelmeergebiet. Rep. VI Intern. Congr. on Quaternary (Warsaw 1961) Geomrph. Sec., **3**: 127–148.

Kaiser, K. (1969) The climate of Europe during the Quaternary Ice Age. Quaternary Geology and Climate. Vol. 16 of the Proceedings of VII Congress of INQUA. 10–37.

Kaiser, K., Kempf, E. K., Leroi-Gourhan, Arl., und Schütt, H. (1973) Quartärstratigraphische Untersuchungen aus dem Damaskus-Becken und seiner Umgebung. Zeit. Geomorph., **17**: 263–353.

King, L. C. (1968) Pediplanation. Fairbridge ed. Encyclopedia of Geomorphology, 818–820.

van Liere, W. J. (1960–61) Observation on the Quaternary of Syria. Publication of the general directorate Antiquities and Museum in the Syrian Arab Republic, 7–69.

Mensching, H. (1958) Glacis- Fussfläche- Pediment. Zeit. Geomorph., **2**: 165–186.

Mouterde, P. (1970) Nouvelle Flore du Liban et de la Syrie. Dar el-Machreq Editeurs. Beyrouth.

Ollier, C. D. (1969) Weathering. 304 pp. Oliver & Boyd. Edinburg.

Powers, R. W., Ramirez, L. F., Redmond, C. D., and Elberg, E. L. (1966) Geology of the Arabian Peninsula. U.S. Geol. Surv. Prof. Pap. 560-D. 1–147.

Rubin, M., Likens, R. C., and Berry, E. G. (1963) On the Validity of Radiocarbon Dates from Snail Shells. J. Geol., **71**: 84–89.

Rohdenburg, H. (1969) Hangpedimentation und Klimawechsel als wichtigste Faktoren der Flächen- und Stufenbildung in den wechselfeuchten Tropen an Beispielen aus Westafrika, besonders aus dem Schchtstufenland Südost-Nigerias. Gies. Geogr. Schr. H. 20. 57–152.

Suzuki, H. and Takai, F. (eds.) (1973) The Palaeolithic Site at Douara Cave. Part I. Univ. Museum, Univ. Tokyo, Bull. 5.

Twidale, C. R. (1967) Origin of the piedmont angle as evidenced in South Australia. J. Geol., **75**: 393–411.

Twidale, C. R. (1968) Pediments. Fairbridge ed. Encyclopedia of Geomorphology. 817–818.

de Vaumas, E. (1970) Phénoménes cryogéniques de la cote Libanaise. Rev. Géogr. Phy. Geol. Dynam., **XII**: 265–292.

Wirth, E. (1958) Morphologische und bodenkundliche Beobachtungen in der Syrisch-Irakischen Wüste. Erdk. **XII**: 26–42.

Wolfart, R. (1967) Geologie von Syrien und dem Libanon. 326 pp. Gebrüder Borntraeger, Berlin.

Zohary, M. (1973) Geobotanical Foundations of the Middle East. 739 pp. Gustav Fischer Verlag. Stuttgart.

CHAPTER 4

Stratigraphy and Paleoenvironments of the Deposits in and around the Douara Cave Site

Kunihiko Endo

Department of Applied Earth Sciences, College of Humanities and Sciences, Nihon University

4. 1 INTRODUCTION

The Douara Cave Site, Mugharet ad Douara I, is situated on the southern slope of Jabal ad Douara (Douara Mountain), made of Paleogene limestone and marl (Ponikarov, et al., 1966), bordering the northern margin of the Palmyra basin in the central part of Syria. Many alluvial fans formed on the southern foot of the mountain slope down toward the central marshland (Sabkhet) in the Palmyra Basin. The altitude of the cave is about 550 meters above sea level, about 170 meters above the plain of the central marshland and about 20 meters above the top of the alluvial fan surfaces. The Palmyra Basin lies in the area of desert or sub-desert, and so the Douara Cave, as well as the sites of Jerf Ajla (Schroeder, 1969; Goldberg, 1969) and Yabrud (Rust, 1950; Farrand, 1965, 1972) in Syria, is one of the sites located in the most arid regions of any excavated Middle Paleolithic site. The 1974 season's excavation was made at the Douara Cave site and at three test trenches of the open sites (WF-1, 2 and 3) on the alluvial fan horizontally 50 to 200 m downwards from the cave (Fig. 4. 1). In this chapter, the writer will attempt to discuss the stratigraphical and paleoenvironmental relationships between the Douara Cave site and the alluvial fan deposits and some other deposits around the cave.

4. 2 STRATIGRAPHY OF THE DOUARA CAVE DEPOSITS

The 1970 season's preliminary excavation of the Douara Cave site made it clear that the deposits are more than 4. 3 m in thickness, and are composed of 14 stratigraphical layers (Akazawa, Baba and Endo, 1973). During the 1974 season, the upper part of the deposits in the Douara Cave was excavated from the surface to 1–1.5 m in depth. The excavated area covered about half of the cave.

The deposits excavated in this season can be divided into several layers according to differences in various sedimentological features such as color, texture, hardness, some sedimentary structures, boundaries of adjoining layers, and so on. It was revealed on referring to the cultural remains that the main lithostratigraphical boundaries correspond to those between the cultural assemblages. Consequently, based on the lithostratigraphical characteristics and the differences in the assemblages of cultural remains, four stratigraphical divisions were determined. They are the First, the Second, the Third and the Fourth Horizons in descending order. The Second to the Fourth Horizons are further subdivided lithostratigraphically (Table 4. 1).

The dating of the deposits through the ^{14}C Method was carried out by Dr. W. G.

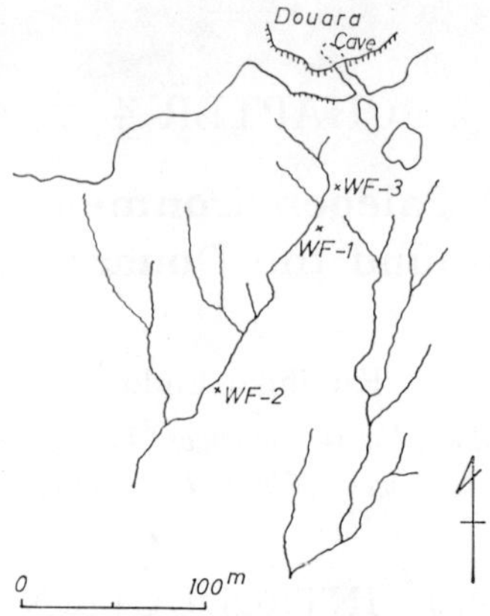

Fig. 4.1. Locality map of the Douara Cave and the test trenches of WF-1, 2 and 3.

Mook, Groningen University, and by Dr. Kobayashi, the University of Tokyo, while dating by the fission track method was carried out by Dr. Suzuki, Rikkyo University.

4.2.1 The First Horizon (I)

The First Horizon is subdivided into a felt-like layer, a blackish-brown rubble layer, and a dark brown silty rubble layer.

The felt-like layer, the uppermost one, is about 1 to 4 cm thick, and is composed of goat and sheep feces and fine materials. It covers the whole of the earth floor of the cave, which is flat and inclined slightly toward the entrance.

The blackish-brown rubble layer is 0 to 20 cm in thickness, and is composed of limestone rubble, sand, and silt. The matrix is mostly loose, porous, and rich in humus. The boundary between this layer and the underlying layers is wavy.

The dark brown silty rubble layer contains a large amount of angular limestone rubble with a slightly more compact matrix than that of the overlying layers. The limestone rubble is mainly pebble to cobble size, with a few pieces of boulder size. The matrix is composed mainly of sand and silt and is very granular. The thickness varies

Table 4.1. Stratigraphy of the Douara Cave Deposits

	1974 season		1970 season
Neolithic to Present	First Horizon		Surface
Epi-Paleolithic	Second Horizon	Unit A	Layer A
		Unit B	Layer B
Middle Paleolithic	Third Horizon	Unit A	Layer C
		Unit B }	Layer D
	Fourth Horizon	Unit A }	
		Unit B }	Layer E
			Layer F
			Layer G

from 0 to 40 cm, from place to place. The deposits are thicker in small pocketlike depressions which were dug into the underlying layers, such as that at the interior end of the cave (Fig. 4. 4, section E-F).

The First Horizon contains flint tools of the Middle Paleolithic and the Neolithic ages, and some potsherds and bronze tools which are probably of the Islamic age. This fact confirms that the deposits of the First Horizon are inhomogeneous and formed mainly of reworked materials.

4. 2. 2 *The Second Horizon (II)*

The Second Horizon is a yellowish to pinkish-gray color, about 110 cm in thickness, loose with a poor matrix, and characteristically contains much fragmental limestone rubble. In the excavated area during the 1974 season, the distribution of the Second Horizon is restricted to the area of grids 8-10 and 8-11 near the entrance of the cave. As shown in Fig. 4. 6, the distribution and the thickness of the Second Horizon are reconstructed on the basis of the data obtained during the 1974 excavation and Trench A and B in the 1970 season. The depression filled up by the Second Horizon runs in the direction of SW-NE, having its center in the grid 7-10 of Trench B 1970 season.

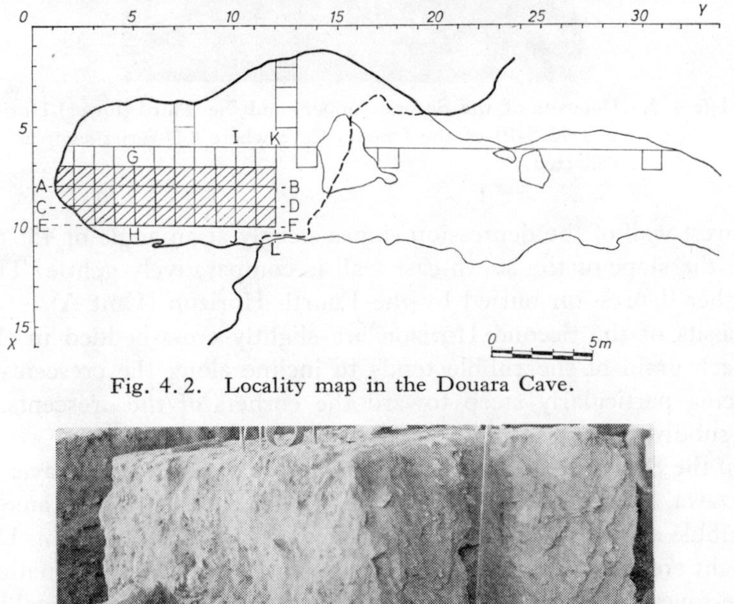

Fig. 4. 2. Locality map in the Douara Cave.

Plate 4. 1. Deposits in grid 8–10 of the Douara Cave.

Plate 4. 2. Deposits of the Second (upper) and the Third (lower) Horizons in grid 8-10 of the Douara Cave where CD-series samples were collected.

The north-west wall of the depression slopes steeply at an angle of 45° to 80°. On the other hand, the slope of the south-east wall is comparatively gentle. This depression lies on another depression buried by the Fourth Horizon (Unit A).

The deposits of the Second Horizon are slightly cross-bedded in the shape of a crescent. Each grain of the rubble tends to incline along the crescent-shaped cross-bedding, being particularly steep toward the corners of the crescents. The Second Horizon is subdivided into Units A and B.

Unit A of the Second Horizon (II A), which was designated as Layer A in the 1970 season (Akazawa, Baba and Endo, 1973; Endo, 1973), includes a large amount of angular limestone rubble and silty sand. Its thickness ranges from 20 to 50 cm. Unit A consists of six to eight crescent-shaped beds, with fine and coarse ones alternating. Finer beds contain limestone rubble with a frequency of less than 40% of the total in weight, and coarser ones with a frequency of 50 to 60%. The upper half of Unit A consists of yellowish-gray to yellowish-brown sandy rubble layers, having a maximum thickness of about 45 cm in the grid 8-10. It is loose and porous. As shown in Fig. 4. 5b, a large fallen rock is included in it. The lower half of Unit A consists of pinkish-gray sandy rubble layers and is slightly more compact, compared to the upper half. It has a maximum thickness of about 60 cm, and consists of three to five alternating beds.

Unit B of the Second Horizon (II B), having been designated as Layer B in the 1970 season (Akazawa, Baba and Endo, 1973; Endo, 1973), consists of a rather loose, light brown sediment having a maximum thickness of about 25 cm in the grid 8-10. Its

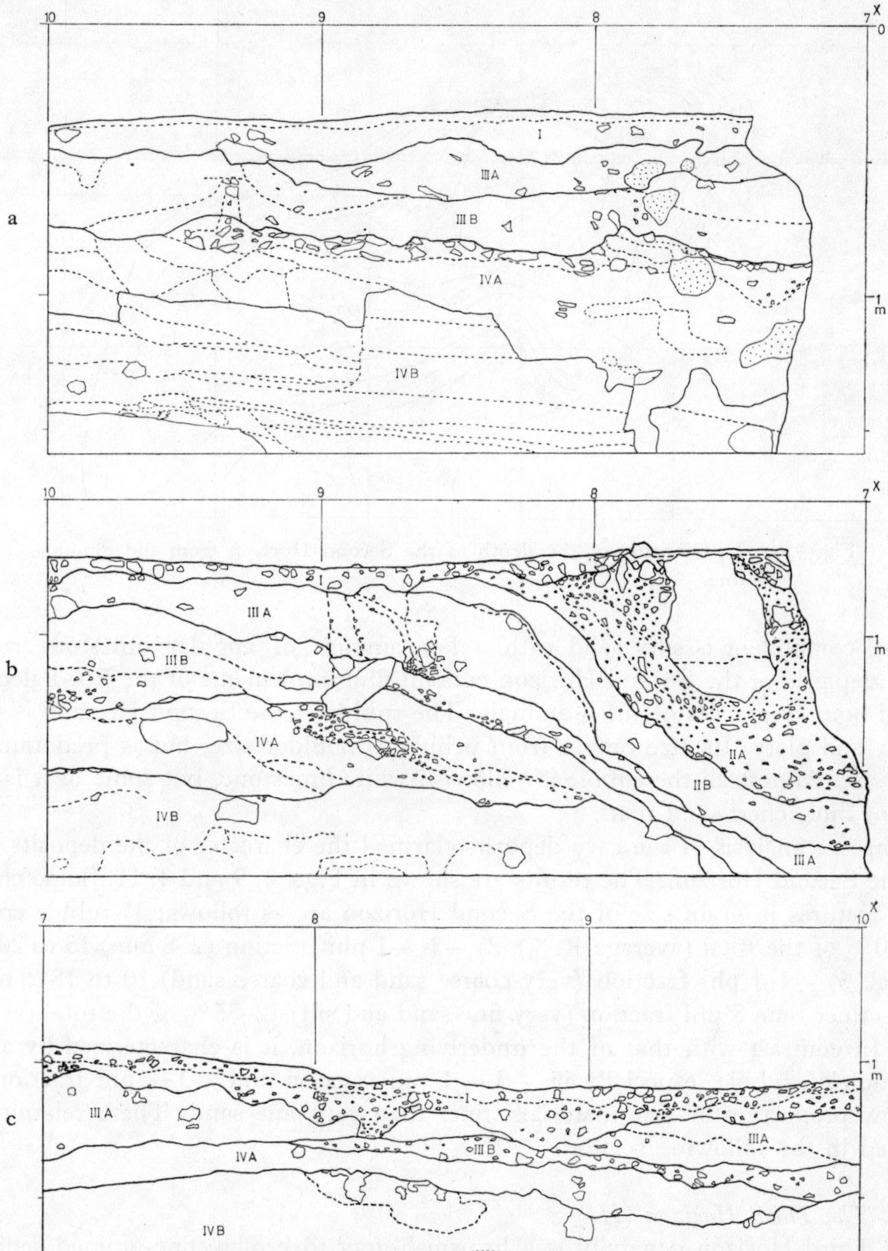

Fig. 4. 5. Vertical sections of the Douara Cave deposits along the transversal lines of Y=5, 10 and 12 (Localities are shown in Fig. 4. 2).
a. Y=5 line (G-H), b. Y=10 line (I-J), c. Y=12 line (K-L).

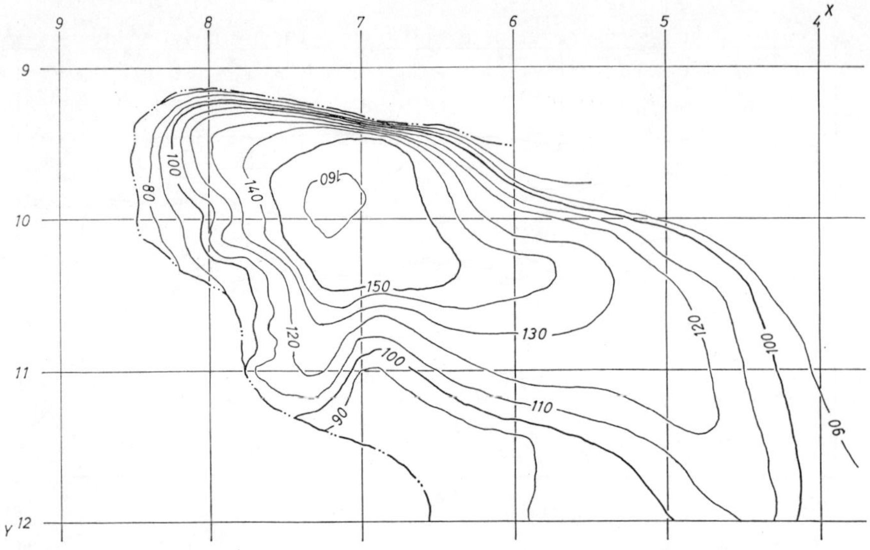

Fig. 4. 6. Distribution of the depth of the Second Horizon from the datum line.

principal component is silty sand with a fair amount of angular limestone rubble.

The deposits of the Second Horizon contain flint implements of the Epi-Paleolithic age and bone fragments of small animals. The rubble of the Second Horizon is fresh, angular, and platy. Its size ranges from pebble to boulder size, but is predominantly pebble size. Almost all the rubble is yellowish-white limestone, but some of it is fragments of flint, chert, and marl.

Grain-size analysis of the cave deposits clarified the character of the deposits taken from the Second Horizon. The results are shown in Figs. 4. 9 and 4. 11. Some characteristic features in grain size of the Second Horizon are as follows: 1) rubble content 30 to 60% of the total (average 40%), 2) —3-—1 phi fraction (2–8 mm) 15 to 20% of the total, 3) —1–1 phi fraction (very coarse sand and coarse sand) 10 to 18% of the total, 4) finer than 3 phi fraction (very fine sand and silt) 12–35% of the total (average 20%). In contrast with that of the underlying horizon, it is characterized by a high proportion of rubble, especially in —3-—1 phi fraction and —1–1 phi fraction and by a low proportion of fine materials, finer than very fine sand. These relations are discussed in the following section.

4. 2. 3 The Third Horizon (III)

The Third Horizon generally is a brownish-gray to brown, fine-grained sediment composed mainly of sand and silt, and is intercalated by some rubble and ash layers in places. It is distributed throughout the excavated area and is 20 to 80 cm in thickness. A large number of flint implements of the Middle Paleolithic Age and animal bones are found in this horizon. As a whole, it corresponds to the deposits designated as Layers C and D in the 1970 season. The Third Horizon is subdivided into two Units, A and B.

Unit A of the Third Horizon (III A), in the interior from the Y=9 line, is a pale brownish-gray to pinkish-gray sediment and covers the underlying layers horizontally except at the interior end of the cave. It is more compact than the overlying First Horizon which is the most loose sediment. It is 10 to 50 cm and averages 20 cm in thickness. The upper half of Unit A contains a large amount of limestone rubble in pebble to cobble size and sand-sized materials mainly finer than 3 phi. The lower half contains, in general, as its main component sand and silt with a fair amount of limestone rubble (Fig. 4. 10). A small number of lenses composed of charcoal, ash, or humic soil are intercalated in the upper half of Unit A. This unit corresponds to the upper part of Layer D defined in the 1970 season.

In the exterior of the cave, in the grids between Y=8 and Y=11, a hard, whitish-gray calcareous concretion layer is found. This massive and homogeneous calcareous concretion, as a whole, makes up Unit A of the Third Horizon. It is the same layer that was designated as Layer C in the 1970 season (Akazawa et al., 1973). It is 25 to 40 cm in thickness and covers with a nearly constant thickness the concave slope of the depression which was cut into the underlying layers. Its principal components in the area are limestone rubble and the fraction finer than 3 phi. The proportion of rubble is 20 to 50% and it averages 30% of the total. That of the fraction finer than 3 phi is 30 to 50% of the total.

A great number of the Middle Paleolithic flint implements contained in Unit A are of the assemblage of the Levantine Mousterian in character. Traces of animal burrows dug in this unit are found in abundance near the entrance of the cave, so that the slight disturbance of the sediment probably caused a little admixture of non-Paleolithic implements. A small number of ostrich eggshell fragments is included.

Unit B of the Third Horizon (III B) consists of a brown sediment distributed throughout the excavated area. The thickness varies from 10 to 60 cm and is generally about 30 cm. In the interior of the cave, the bottom of the unit descends gradually inward, and at the innermost part, grids 8-1 and 8-2, it attains a maximum depth of 1.25 m below the floor surface. The boundary with the overlying Unit A is not very clear. The sediment is not hard but compact, and more or less sticky. In general, it is fine-grained,

Plate 4. 3. Rubble of the Second Horizon, the Douara Cave.

Plate 4. 4. Rubble of Unit A of the Fourth Horizon, the Douara Cave.

containing a lot of very fine sand fraction as the dominant component. In places, it includes abundant limestone rubble, but this is less than 30% of the total. The limestone rubble is mostly whitish colored.

In grid 8-1, the innermost part of the cave, a narrow hollow was found inside the sediment of Unit B at a depth of 110 cm. It is about 30 to 40 cm high and about 100 cm in width. The ceiling and the sidewalls of the hollow are of limestone. The floor of the hollow is composed of the sediment of Unit B, sticky silty sand with pebble-sized white limestone rubble, and the whole surface covered by a plate of travertine.

Unit B of the Third Horizon contains a large number of flint implements of the Levantine Mousterian type similar to those of Unit A (B. Endo et al., Chapter 5). There is also a large number of animal bone fragments, and eggshell fragments of ostrich, *Struthio* camelus. A small number of bola stones of barite are also included.

4. 2. 4 The Fourth Horizon (IV)

The Fourth Horizon, as a whole, is a deep brown to dark brown, fine-grained sediment, accompanied by a large number of intercalations of whitish ash layers and by remarkable rubble layers. It is the lowest horizon excavated in 1974, and is distributed throughout the excavated area with a maximum thickness of some 85 cm. This Horizon contains a large number of flint implements of the Middle Paleolithic age and animal bones. It corresponds approximately to the deposits designated as Layer E in the 1970 season. This horizon is subdivided into two Units, A and B.

Unit A of the Fourth Horizon (IV A) is composed of an upper, remarkable rubble layer and a lower, dark brown, sandy silt layer. It is distributed throughout the excavated area, and is 10 to 45 cm in thickness. In places, Unit A is truncated by the overlying layers. The bottom is rugged and descends gradually inward from the center of the cave.

The upper rubble layer shown typically on the A-B vertical section (Fig. 4. 4) consists of a large amount of whitish limestone rubble in cobble to pebble size and calcareous concretion. Typically, the calcareous concretion is developed conspicuously along the top of the rubble layer so that the rubble is cemented tightly. Its boundary with Unit B of the Third Horizon is very clear because of this calcareous concretion, but it is occasionally obscure where the calcareous concretion and the rubble layer were truncated.

The lower silty sand layer consists of 55 to 80% sand-sized materials, and includes 5 to 20% silt-sized materials plus 15 to 25% limestone rubble as the additional components. The limestone rubble is whitish colored, well weathered and chiefly fine pebble size. This layer is sticky in the interior of the cave, while, in the exterior of the cave it is porous but tight. The lower portion of the layer is especially humic and rather porous because of the abundance of granule- to coarse sand-sized materials. In places, the lowermost part of the layer is rich in limestone rubble. As shown in the E-F vertical section of Fig. 4. 4, cross stratifications dipping outward are conspicuous particularly between the Y=4 line and the Y=9 line. The upper rubble layer is deposited continuously on the lower layer or changed gradually from the lower layer.

Unit A contains a large number of flint implements of the Middle Paleolithic age, which is different in character from those of the Third Horizon (B. Endo et al., Chapter 5). In addition, it includes numerous animal bone fragments and a small amount of ostrich eggshell fragments.

Unit B of the Fourth Horizon (IV B), inside the Y=8 line, is characterized by the thick alternation between whitish to grayish ashy layers and blackish charcoal lenses. This alternation has a maximum thickness of 60 cm around the center of the cave. It is distributed mostly throughout the inner portion of the cave in the shape of an ellipse with its center around the grid 8-5. Each ashy layer and charcoal lens dips to the center and so the floor after the 1974 excavation at a depth of 160 cm below the datum line showed an elliptic horizontal section (Plate 4. 6). It is truncated at the interior end of the cave and near the margin of the ellipse. The base of the alternation is elevated toward the interior end. In grids 8-2, 8-3, 9-3, and 9-4, the thick charcoal lenses dipping outward are deposited to form a cover over an arrangement of limestone rubble

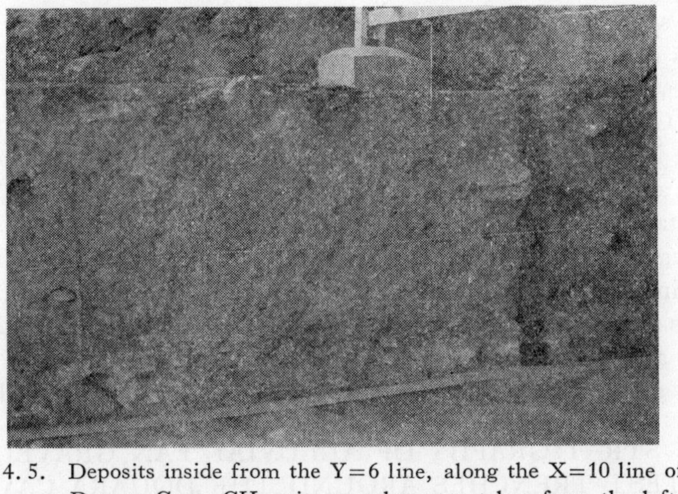

Plate 4. 5. Deposits inside from the Y=6 line, along the X=10 line of the Douara Cave. CH-series samples were taken from the left side vertically.

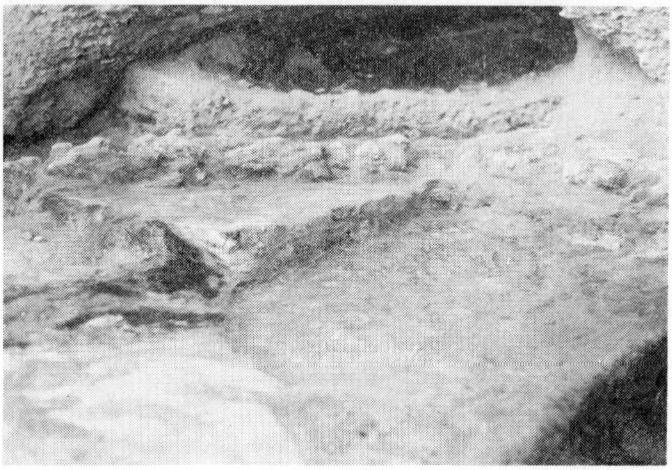

Plate 4. 6. Arrangement of stones bordering the hearth and the hollow at the innermost part of the Douara Cave. The floor of the hollow is covered with a sheet of travertine.

and a hard, rugged surface of concretion. The limestone rubble is cobble to boulder size and reddish-yellow or blackish colored. These facts indicate that Unit B of the Fourth Horizon is composed mainly of hearth deposits.

Outside the Y=9 line, Unit B is a homogeneous and very hard sediment. It has a maximum thickness of 50 cm. It is gray to dark gray in color, and rich in fine organic materials. A small degree of stratification due to the intercalations of very hard concretion layers is found.

Unit B includes a sand-sized fraction of 50 to 85% as the predominant component. In some places it contains a large amount of weathered limestone rubble, whitish colored. The predominance of the sand-sized fraction is closely related to the abundant occurrence of amorphous silica fragments from this unit (Endo, 1973). These fragments are porous and of spongy structures and/or with curved surfaces, some of which resemble silica skeletons of plant tissues (Matsutani, 1973).

The samples taken from the hearth were dated by the ^{14}C dating method in the laboratories of Groningen University and Tokyo University. According to these datings, the age for the Middle Paleolithic hearth deposits is more than 43,200 or 52,000 y.B.P., as shown in Table 4. 3.

Unit B contains a large number of flint implements of the Middle Paleolithic age. This cultural assemblage is similar to that of Unit A, but obviously different from that of the Levantine Mousterian in the Third Horizon. The flint implements from the hearth deposits are, in general, patinated bluish or blackish-gray colored and are fragile, presumably as a result of undergoing combustion. A small number of ostrich eggshell fragments and a large number of animal bone fragments are also included.

4. 3 THE STRATIGRAPHY OF ALLUVIAL FAN GRAVELS IN THE TEST TRENCHES AROUND THE DOUARA CAVE

In front of the Douara Cave, alluvial fans extend over toward the center of the

Palmyra Basin. Veneers of talus deposit cover the southern foot of the Douara Mountains, and change gradually into alluvial fan deposits toward the center. A great number of limestone blocks and boulders fallen from the southern slopes of the Douara Mountains are scattered sporadically on the steeper surfaces of the taluses. The surfaces of the alluvial fans and taluses are dissected vertically and linearly by a dense network of wadis. Within a distance of some 200 m from the northern margin of the basin, the wadi floors are narrow, 0.5 to 1.5 m wide and 1 to 2.5 m beneath the fan surfaces. In some places a narrow terrace is distributed along the wadis, undercutting the alluvial fan deposits. It has a height of 0.5 to 1 m from the wadi floor.

A vast number of flint implements of the Neolithic and the Paleolithic ages are found on the surface of the talus and the fan in front of the Douara Cave. Three test ternches,

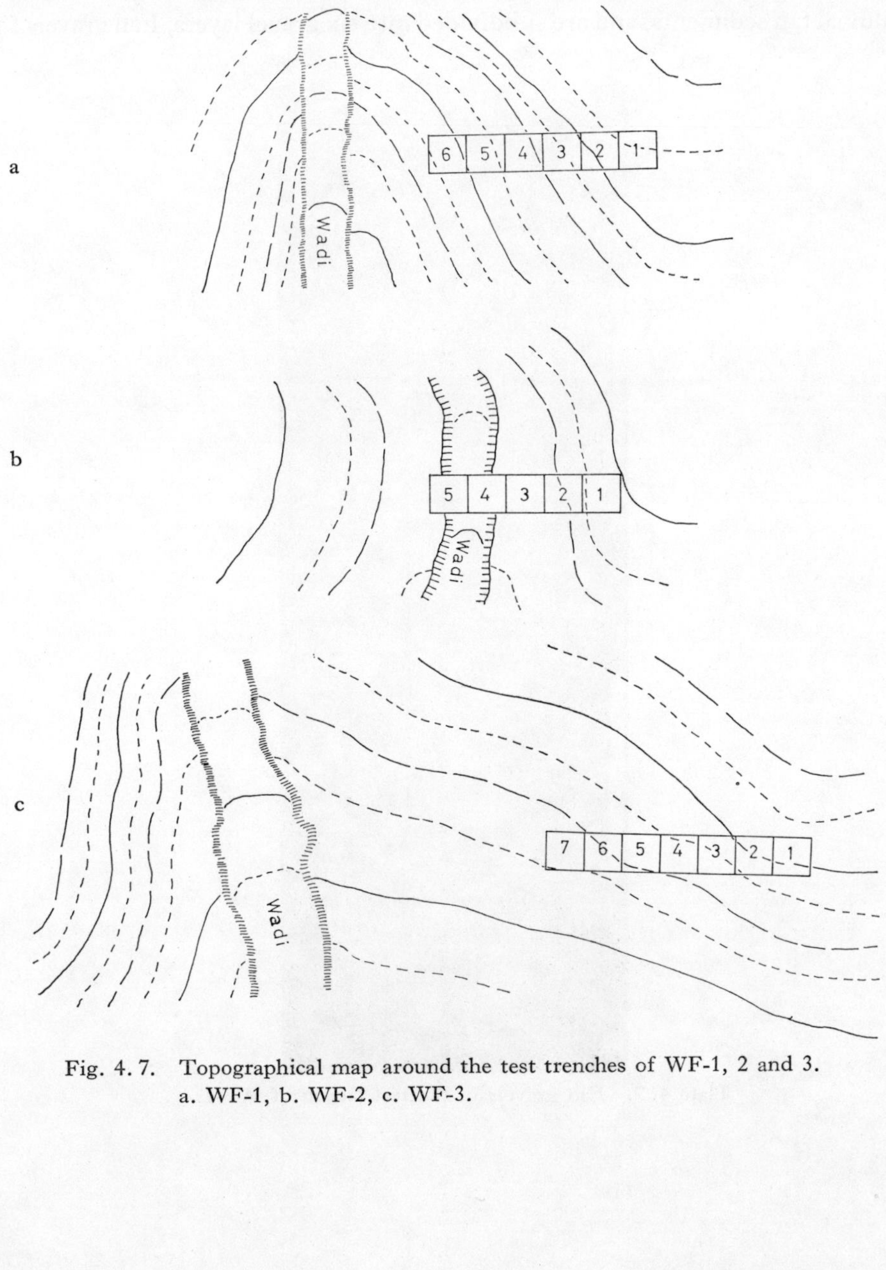

Fig. 4. 7. Topographical map around the test trenches of WF-1, 2 and 3.
a. WF-1, b. WF-2, c. WF-3.

WF-1, 2 and 3, were excavated on the alluvial fan (Figs. 4. 7 and 4. 8). Test trench WF-3 is located about 70 m south of the entrance of the Douara Cave (Fig. 4. 1). Flint implements were found most abundantly in the deposits of WF-3. Test trench WF-1 is located about 100 m south of the Cave. Test trench WF-2 is located about 200 m south of the cave. Trench WF-2 has the smallest number of flint implements among the three trenches.

4. 3. 1 WF-3 test trench

The deposits of the WF-3 test trench have a thickness of about 3.2 m overlying the undulating basement of limestone. The cross-section for trench WF-3 shows the most typical profile of the three trenches (Figs. 4. 7c and 4. 8c). The deposits of WF-3 are composed of a great amount of limestone gravel and a matrix of sandy materials. They form alluvial fan sediments, and are subdivided into six gravel layers, Fan gravels I to VI.

Plate 4. 7. Fan gravels in the test trench of WF-2.

Fan gravel I, the uppermost one, is a loose and pale grayish-brown sediment containing a large amount of limestone gravel in pebble to cobble size, being especially rich in cobbles. Most of the gravel is angular or subangular in shape and the occurrence of cobble-sized gravel is rather sporadic and irregular. Fan gravel I includes the greatest number of flint implements of the Pre-Pottery Neolithic age (PPN) among the layers and a small number of flint implements of the Middle Paleolithic age (B. Endo et al., Chapter 5).

Fan gravel II is a more or less compact, pale brown sediment having a maximum thickness of about 100 cm in grid 1 which is the furthest from the present valley or wadi. The distribution of Fan gravel II is restricted to grids 1, 2 and 3, where the former valley indicated by the lowering of the basement is located. Fan gravel II consists mainly of a vast amount of limestone gravel, chiefly of pebble size and angular to

Table 4.2. Stratigraphy of the Deposits in the Test Trenches on the Alluvial Fan

		WF-1	WF-2	WF-3
Present		Present gravel	Present gravel	
Neolithic	Pottery	Terrace gravel	Terrace gravel	
	PPN	Fan gravel I Fan gravel II	Fan gravel I Fan gravel II	Fan gravel I Fan gravel II
Epi-Paleolithic				
Middle Paleolithic		Fan gravel III Fan gravel IV	Fan gravel III Fan gravel IV	Fan gravel III Fan gravel IV
?				Fan gravel V Fan gravel VI
				Basement of limestone

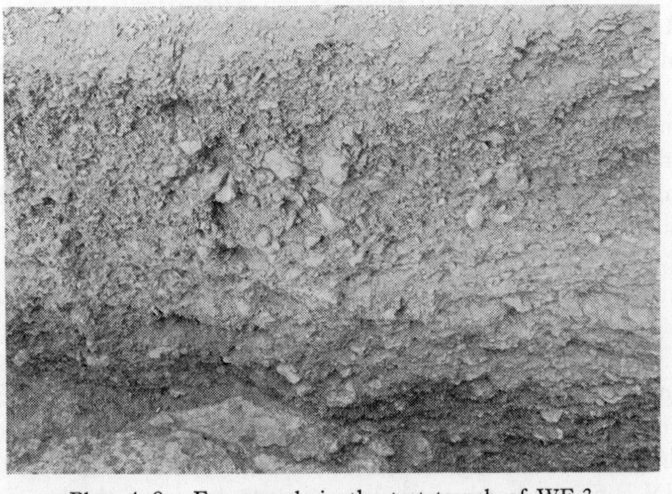

Plate 4.8. Fan gravels in the test trench of WF-3.

Plate 4. 9. Basement of the test trench of WF-3.

subangular in shape. The upper half of the layer lacks a matrix of sand or muddy materials, but the lower half has a fair amount of sandy matrix and a deeper brown color than the upper. Slight stratifications are recognizable, mainly in the upper half. A small amount of limestone gravel of cobble size is found in the lower half but only sporadically. Along the surface of the lower half, gravel is slightly cemented by a whitish, calcareous crust. Fan gravel II includes a large number of flint implements of the Pre-Pottery Neolithic and the Middle Paleolithic age with a small number of Epi-Paleolithic flint implements (B. Endo et al., Chapter 5).

Fan gravel III is a tight and massive sediment, having a thickness of 40 to 130 cm. It is distributed throughout the trench. The bottom of the layer descends to the former valley. The color of Fan gravel III is deep brown in the upper part of the layer, pinkish-brown in the middle and a pale reddish-brown in the lower. In general, a fair amount of limestone gravel of cobble size is included sporadically. In the middle part of the layer a large amount of limestone gravel of pebble size is contained. The matrix of the layer is dominant in general and sticky, consisting of sand and silt-size materials. The stratification is very poor as a whole, but in the middle part it is recognized slightly. A considerable number of whitish spots of calcareous crust are scattered in the lower part of the layer. Fan geavel III includes a large number of flint implements of the Middle Paleolithic age. Almost all of the Middle Paleolithic flint implements correspond in character to those from Units A and B of the Third Horizon in the Douara Cave Site. Only a small number of the flint implements correspond in character to those from the Fourth Horizon (B. Endo et al., Chapter 5).

Fan gravel IV is a very hard, gravelly sediment, having a thickness of 10 to 50 cm.

It is distributed throughout the trench and dips to the east on the vertical section, namely toward the former valley. It is deep brown in color with plenty of whitish spots of calcareous crust. Its principal component is a vast amount of limestone gravel in pebble to cobble size, chiefly in pebble size. Each piece of gravel is angular to subangular and comparatively flat in shape. The matrix is poor but composed mainly of granule to fine pebble-sized materials. The stratification is remakably distinct as a whole. Fan gravel IV includes a fair number of flint implements of the Middle Paleolithic age, whose character corresponds to those from Units A and B of the Third Horizon in the Douara Cave Site (B. Endo et al., Chapter 5).

Fan gravel V is a very hard and massive sediment, having a maximum thickness of 50 cm in the location of the former valley, but it seems to be truncated partly by the overlying layer. A fair amount of limestone gravel of cobble size is included sporadically in the dominant matrix which is very sticky and reddish-brown in color, with whitish patches of calcareous crust. No stratification is recognizable. Probably no flint implements were found in this layer.

Fan gravel VI is the most consolidated sediment directly overlying the basement limestone. It has a maximum thickness of 60 cm at grids 3 and 4 in the former valley. Its color is reddish-brown with a great number of whitish patches of calcareous crust. It is composed mainly of angular limestone gravel in cobble and pebble size. Each piece of gravel is fragmental and flat in shape, and cemented tightly to one another, with a fair amount of matrix of calcareous crust. No flint implements were found in this layer.

4. 3. 2 WF-1 test trench

The test trench of WF-1 is located about 30 m downward from that of WF-3. Deposits with a maximum thickness of about 300 cm were excavated, but the basement was not reached. The deposits are subdivided into six layers, namely Present gravel, Terrace gravel and Fan gravels I to IV (Figs. 4. 7a and 4. 8a).

Present gravel in the trench is a very loose, gravelly sediment which is continuous with the present wadi floor gravel composed of limestone gravel in cobble to boulder size. It is a slope deposit including limestone gravel mainly of pebble size. It contains chiefly a number of flint implements of the Pre-Pottery Neolithic age.

Terrace gravel is a loose and porous sediment cut into the alluvial fan deposits. It has a maximum thickness of 70 cm. It is composed of limestone gravel in pebble to cobble size. It includes a large number of flint implements chiefly of the Pre-Pottery Neolithic.

Fan gravel I, the uppermost layer of the fan deposits, is a loose and porous sediment having a maximum thickness of 50 cm. In the grids near the wadi it is truncated by the Terrace gravel. It is pale grayish-brown in color and composed mainly of limestone gravel of pebble size and granulous materials as matrix. A small amount of cobble-sized gravel is contained near the bottom of the layer. The boundary with the underlying Fan gravel II is clear and slightly ragged. Slight stratification is recognized. It includes a large number of flint implements, mainly of the Pre-Pottery Neolithic age (B. Endo et al., Chapter 5). It corresponds to Fan gravel I of WF-3.

Fan gravel II is a comparatively tight sediment having a maximum thickness of 120 cm

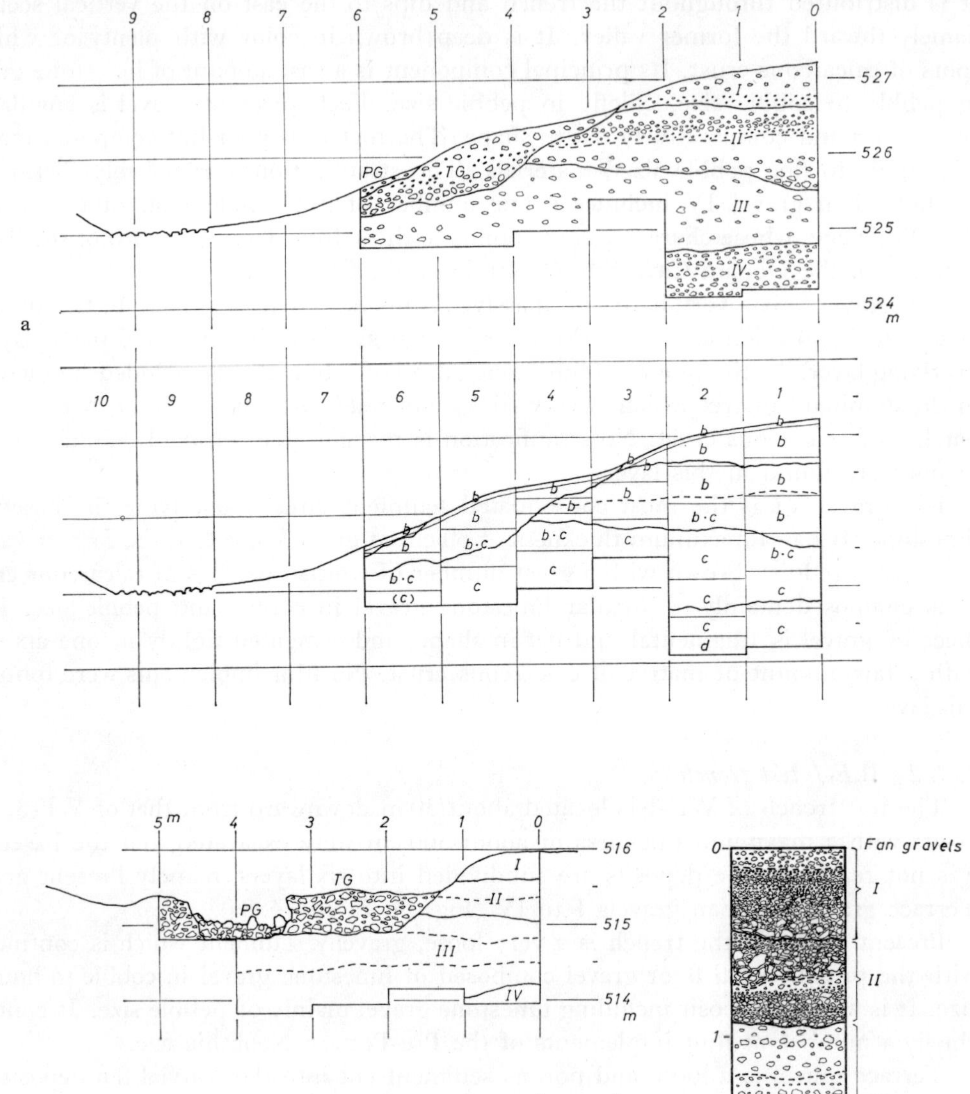

Fig. 4. 8. For explanation see next page.

c

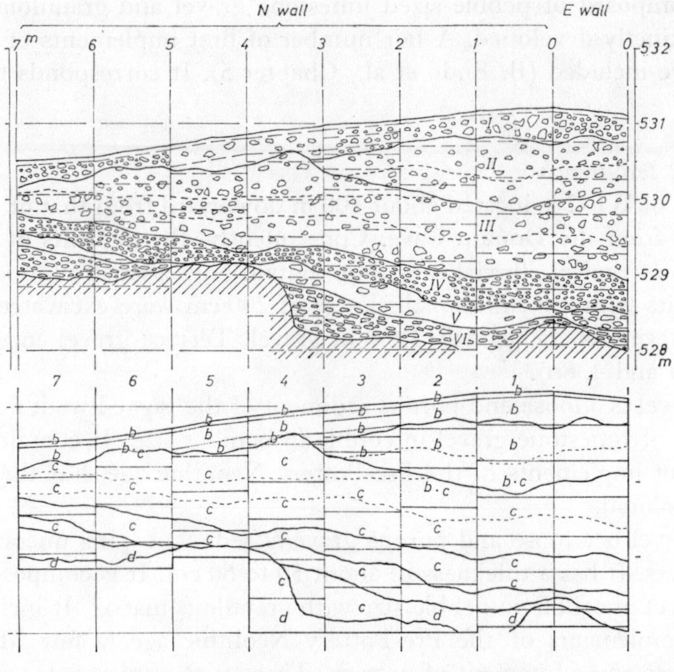

Fig. 4. 8. Vertical sections of the deposits and the cultural remains in the test trenches of WF-1, 2 and 3.
Cultural remains are illustrated according to Akazawa's preliminary results through the partial examination of the collection. Cultural assemblage a: PPN (including Pottery), b: PPN (including a small amount of Epi- and Middle Paleolithic), c: pure Middle Paleolithic, d: no cultural remains.
a. WF-1, b. WF-2, c. WF-3.

in grid 1. It has a brown colored matrix occupying the dominant area. The upper half of the layer contains a large amount of limestone gravel in pebble size. It is stratified slightly. The lower half is a deep brown in color and a thick alternation between massive cobble beds with dominant matrix and thin, granulous beds. Thin calcareous crust occurs in the uppermost parts both of the upper and lower half of the layer. It includes a large number of flint implements of the Pre-Pottery Neolithic and the Middle Paleolithic ages with a few Epi-Paleolithic implements (B. Endo et al., Chapter 5). It corresponds to Fan gravel II of WF-3.

Fan gravel III is a hard and massive sediment having a maximum thickness of more than 100 cm. It is deep brown to reddish-brown in color with a small number of whitish patches of calcareous crust. It is composed mainly of a sticky matrix of fine materials with sporadic occurrence of cobble-, and in some places boulder-sized gravel. Stratification is very poor. It includes a large number of flint implements of the Middle Paleolithic age chiefly corresponding to those from Units A and B of the Third Horizon in the Douara Cave (B. Endo et al., Chapter 5). It corresponds to Fan gravel III of WF-3.

Fan gravel IV is a very hard, gravelly sediment, having a thickness of about 60 cm in grids 1 and 2. It is deep brown in color with numerous whitish patches of calcareous crust. It is composed of pebble-sized limestone gravel and granulous matrix. Stratification is distinctly developed. A fair number of flint implements of the Middle Paleolithic age are included (B. Endo et al., Chapter 5). It corresponds to Fan gravel IV of WF-3.

4. 3. 3 WF-2 test trench

The trench of WF-2 is located about 100 m downward from that of WF-1 and about 200 m distant from the Douara Cave. Consequently, the deposits of WF-2 show the characteristics of typical alluvial fans, but contain only a small number of flint implements. Deposits with a maximum thickness of 200 cm were excavated. They are subdivided into seven layers, namely Present gravel, Terrace gravel and Fan gravels I to V (Figs. 4. 7b and 4. 8b).

Present gravel is a loose and porous sediment of the ragged wadi floor. It includes a large amount of limestone gravel in cobble to boulder size. It contains chiefly a small number of flint implements of the Pre-Pottery Neolithic age, but they are fragmental and whitish-colored.

Terrace gravel is a loose and porous gravelly sediment lying unconformably on the fan gravel layers. It has a thickness of about 50 to 60 cm. It is composed of subangular limestone gravel in cobble to pebble size with granulous matrix. It includes a fair number of flint implements of the Pre-Pottery Neolithic age, a few Middle Paleolithic flint implements and a fragment of pottery. These flint implements are fragmental and whitish colored and have worn-out edges. Consequently, they are probably reworked ones derived from the Fan gravel layers.

Fan gravel I, the uppermost layer of the fan deposits, is a loose and porous sediment with a thickness of about 45 cm. Except in grid 1, it is truncated by the Terrace gravel. It is pale grayish-brown in color and composed mainly of limestone gravel in pebble to granule size. The gravel is subangular to subrounded in shape, sorted considerably, and stratified by pebble beds and granule or fine pebble beds having a thickness of 5 to 10 cm. It includes a large number of flint implements of the Pre-Pottery Neolithic age with a very small number of Middle Paleolithic flint implements. Flint implements of PPN are brownish in color. Those of the Middle Paleolithic age are fragmental. It corresponds to Fan gravel I of WF-3.

Fan gravel II is a comparatively compact sediment having a thickness of about 40 cm. It is distributed only in the grids 1 and 2, and in the other grids it is truncated completely by Terrace gravel. The upper part of the layer is grayish-brown in color and composed of limestone gravel without matrix. The upper half of the upper is granule to pebble size but the lower half is pebble to cobble size with a maximum diameter of 15 cm. The lower part of the layer is brown colored, composed of well-sorted pebbles and granule beds, and well stratified. The gravel is subrounded in shape. Its uppermost portion is massive with dominant matrix which includes whitish patches of calcareous crust. It includes a fair number of flint implements of the Pre-Pottery Neolithic and the Middle Paleolithic ages (B. Endo et al., Chapter 5). It corresponds to Fan gravel II of WF-3.

Fan gravel III is a very hard and massive sediment having a thickness of about 100 cm. It is distributed thickly throughout the trench. The upper 20 to 35 cm of the layer is a deep brown color and composed mainly of a sticky matrix of fine materials with a small amount of limestone gravel. The middle 30 to 50 cm of the layer is slightly reddish-brown in color and includes a large amount of pebble-sized limestone gravel which is cemented by a hard matrix of whitish, calcareous crust. A fair degree of stratification is recognized. At the base of the middle part, a fair amount of cobble-sized gravel is contained. The lower 30 to 50 cm of the layer is reddish-brown in color and composed of a very compact matrix rich in whitish calcareous crust; pebble- to cobble-sized gravel occurs sporadically. Gravel contained in this layer is whitish and well weathered. This layer includes only a small number of Middle Paleolithic flint implements which are patinated and white colored. It corresponds to Fan gravel III of WF-3.

Fan gravel IV is a most tightly consolidated sediment only partly excavated in grid 1. It is deep brown to reddish-brown in color and contains a large amount of weathered,

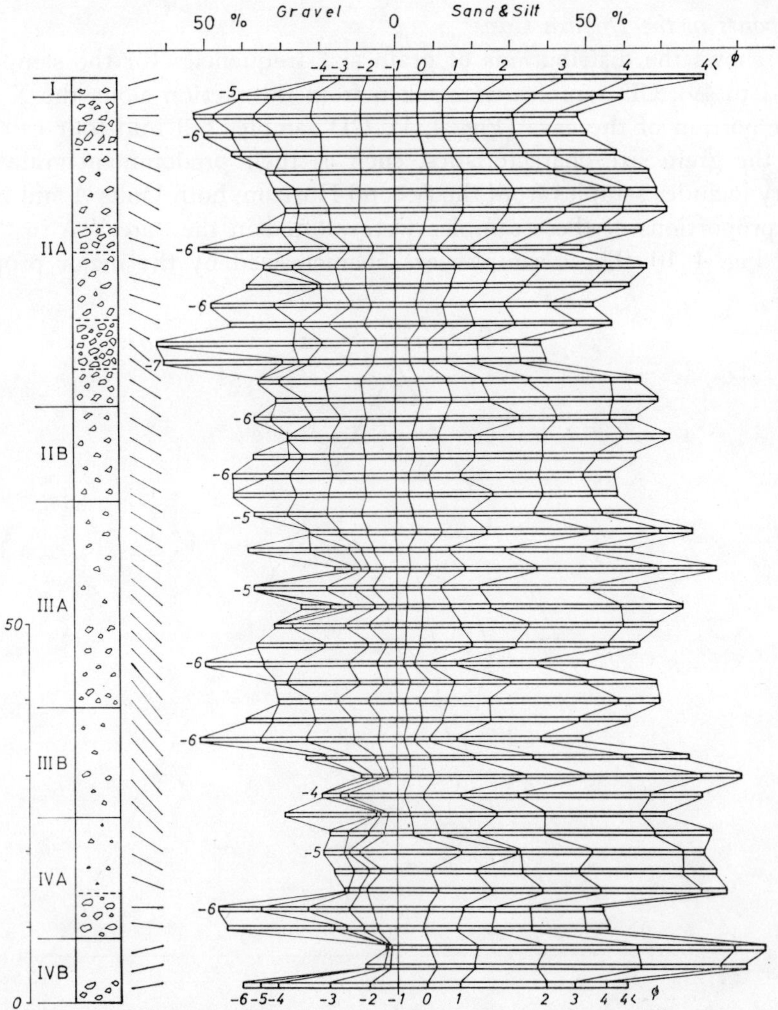

Fig. 4. 9. Grain-size distributions for the CD-series samples from the deposits in the exterior portion of the Douara Cave.

pebble- to cobble-sized limestone gravel cemented tightly together with a matrix of whitish calcareous crust. The small volume of excavation carried out means that we cannot be certain whether it contains some flint implements.

4. 4 GRAIN SIZE OF THE DEPOSITS IN AND AROUND THE DOUARA CAVE SITE

Grain size analyses were done by conventional sieving methods using a ro-tap sieving machine. Size grade intervals used are a quarter phi for the sand and silt fraction and a half or a phi for the gravel fraction. Samples weighing 800 to 2,000 grams for the gravel fraction and samples with 100 to 300 grams for the sand and silt fractions were sieved for ten minutes by the machine.

4. 4. 1 Deposits in the Douara Cave

Fig. 4. 9 shows the distributions of grain size frequencies for the samples of CD series (CD-1 to 49). CD samples were taken from the section along the X=8 line in the exterior portion of the cave (Fig. 4. 4). CD samples fall into four categories according to the grain size characteristics, such as their predominant grain size. The first category includes samples from the Second Horizon, both Units A and B. Gravel: Sand: Silt proportions of these samples were plotted in the "area" A in the ternary diagram of Fig. 4. 10. These samples are characterized by the major proportion of

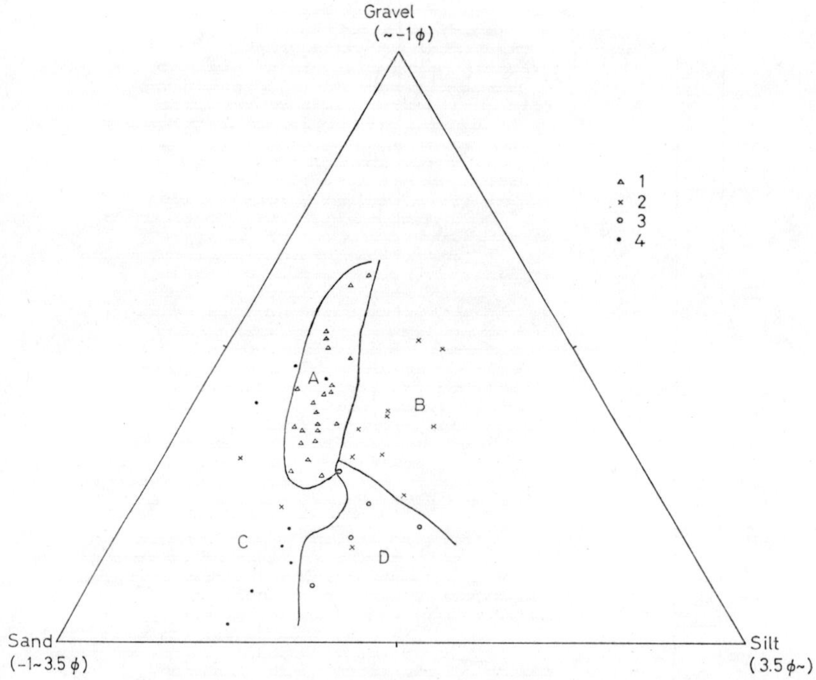

Fig. 4. 10. Ternary diagram showing the proportions of Gravel, Sand and Silt fractions for the CD-series samples from the Douara Cave deposits. 1: Second Horizon, 2: Unit A of Third Horizon, 3: Unit B of Third Horizon, 4: Fourth Horizon.

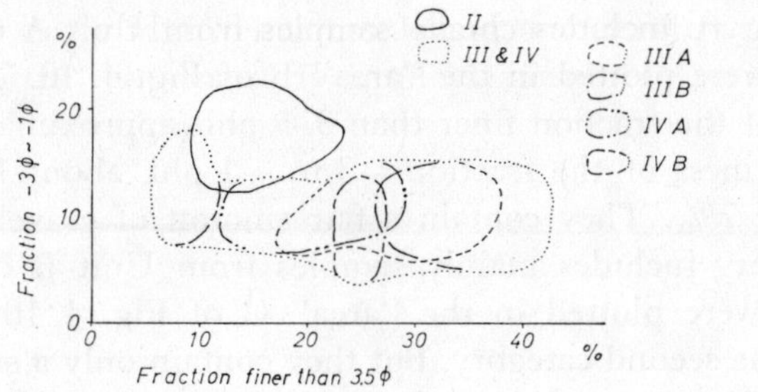

Fig. 4. 11. Fraction −3 phi to −1 phi vs. fraction finer than 3.5 phi for the CD-series samples from the Douara Cave deposits.

gravel-sized rubble, ranging from 30 to 60% and averaging 45% of the total. In particular, they have from 12 to 20% and average 18% of the fraction −3 to −1 phi, fine pebble to granule, which are more dominant, compared with the samples of the other categories. They contain a fair amount of the fraction −1 to 1 phi, namely an average 14% of the total. On the other hand, they contain a comparatively small amount of the fraction finer than 3 phi, 10 to 33% and averaging 24% of the total. The relation of the fraction −3 to −1 phi vs. the fraction finer than 3. 5 phi is shown in Fig. 4. 11. The plotting "area" for the first category is obviously different from that of the other categories.

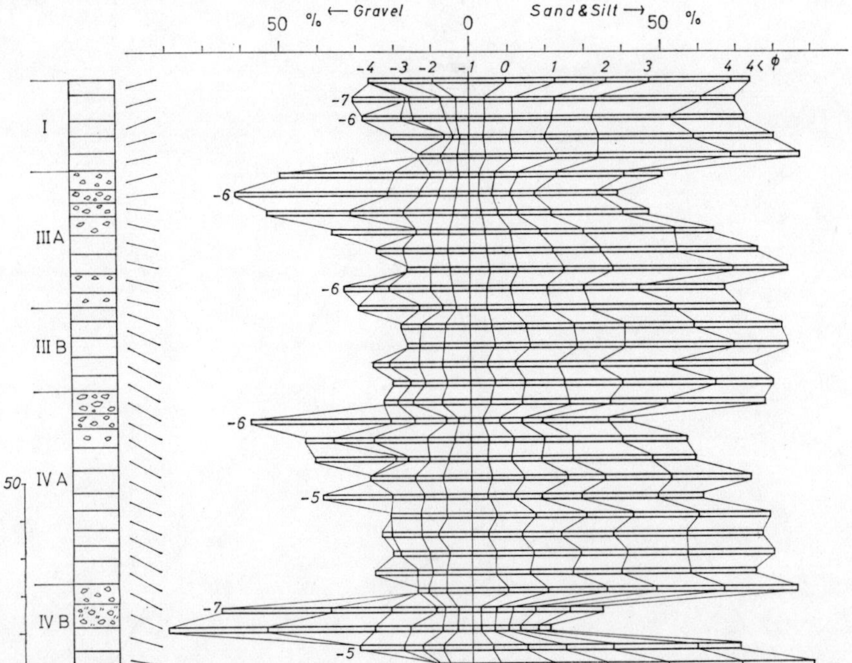

Fig. 4. 12. Grain-size distributions for the CH-series samples from the deposits in the interior portion of the Douara Cave.

The second category includes chiefly samples from Unit A of the Third Horizon. Their proportions were plotted in the "area" B of Fig. 4. 10. They are characterized by the abundance of the fraction finer than 3. 5 phi (approximately coarse silt), about 40%, and the scantiness of the fraction −3 to −1 phi, about 11%, and the fraction −1 to 1 phi, about 7%. They contain a fair amount of gravel-sized rubble.

The third category includes mainly samples from Unit B of the Third Horizon. Their proportions were plotted in the "area" C of Fig. 4. 10. Their characteristics resemble those of the second category, but they contain only a small amount of gravel, averaging 20%, and a large amount of sand-sized materials, about 40%.

The fourth category includes chiefly samples from the Fourth Horizon. Their proportions were plotted in the "area" D of Fig. 4. 10. These samples are divided into two, one having as a main component about 45% gravel-sized materials, and the other being composed mainly of more than 75% sand-sized materials with only 3 to 20% gravel-sized materials. Both are characterized by the predominance of the fraction −1 to 1 phi (very coarse and coarse sand), constituting 15 to 30% and averaging 20%. The fraction finer than 3. 5 phi averages 20% and is less abundant than that of the third category.

Fig. 4. 12 shows the distribution of grain size frequencies for the samples of the CH series (CH-1 to 32). CH-samples were taken from the section along the X=10 line in the interior of the cave (Fig. 4. 4). The lines dividing the four "areas" in Fig. 4. 13 were determined by the lines in Fig. 4. 10. Proportions of almost all the samples were plotted

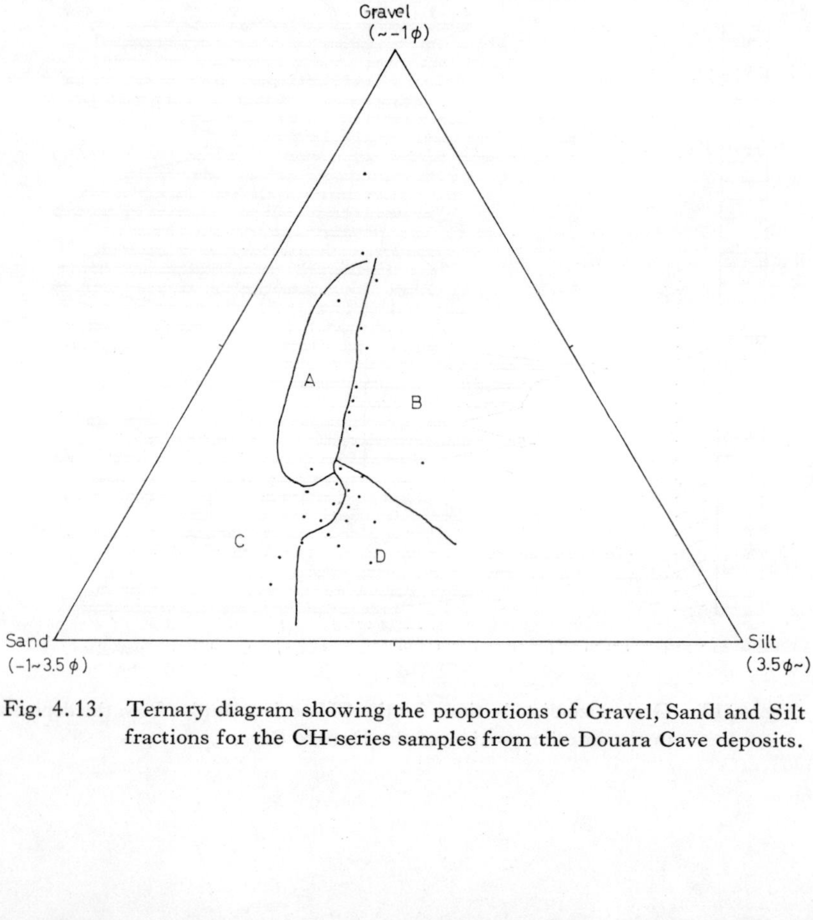

Fig. 4. 13. Ternary diagram showing the proportions of Gravel, Sand and Silt fractions for the CH-series samples from the Douara Cave deposits.

in the "areas" B, C and D, because of the lack of the Second Horizon in the interior of the cave. Each category, excluding the first one, occupies the same "areas" as those of the CD series. Samples from the First Horizon are scattered in the "areas" of the three categories (Fig. 4. 13). It seems to be one of the reasons that the First Horizon is composed mainly fo the reworked materials of the underlying deposits. According to the grain size distribution shown in Figs. 4. 9, 4. 10, 4. 12 and 4. 13, the characteristics of the deposits from each Unit or Horizon are similar between the exterior and the interior portion of the cave.

4. 4. 2 Alluvial fan deposits around the cave

25 samples of the CM series collected in the trench of the WF-2 site were sieved

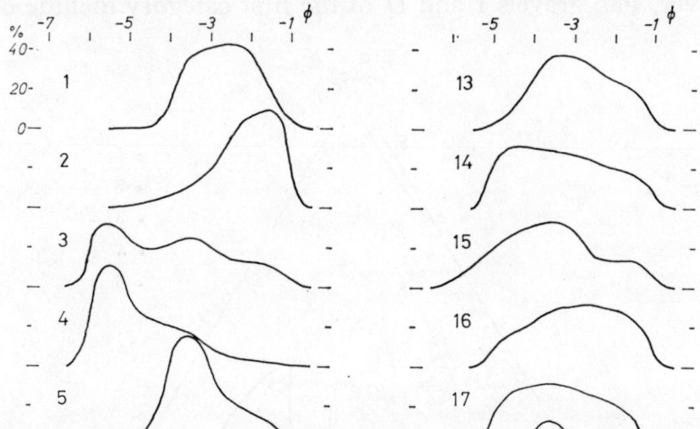

Fig. 4. 14. Grain-size distribution curves for the CM-series samples from the Fan gravels in the WF-2 test trench.
1–7: Fan gravel I, 8–12: Fan gravel II, 13–24: Fan gravel III (13–16: upper, 17–21: middle, 22–24: lower).

using size grade intervals of a phi. Grain size distributions (coarser than −1 phi) for the CM series samples are shown in Fig. 4. 14. These curves fall into two categories. The first one includes curves from Fan gravels I and II. Because of the predominance of graded bedding in Fan gravels I and II, either coarse pebble, fine pebble, or granule is the main component in each bed and so, in general, they give comparatively well-sorted, bell- or spire-shaped curves. However, the subsurface samples of Fan gravel I are not so well-sorted. Grain size frequency curves from Fan gravels III and IV, included in the second category, are dome-shaped ones characterized by poor sorting with the exception of those from the middle Fan gravel III, some of which are as well-sorted as those of the first category. Generally they contain 15 to 30% of the granule-sized fraction, in contrast with the low percentages of this for the samples of the first category. Moreover, Fan gravels I and II of the first category include only about 20%

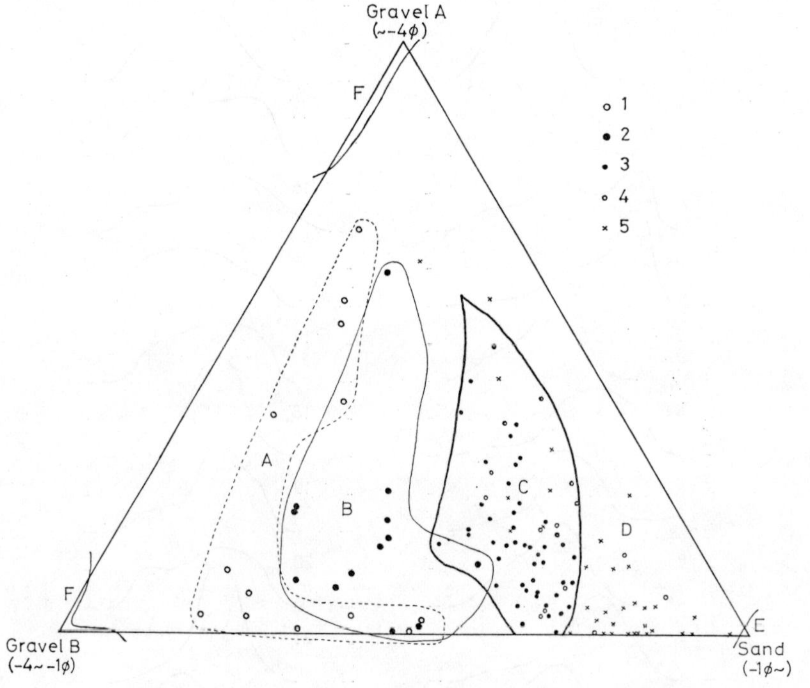

Fig. 4.15. Ternary diagram showing the proportions of coarse gravel (coarser than −4 phi), fine gravel (−4 phi to −1 phi) and sand (finer than −1 phi) for the samples from the Fan gravels in the WF-2 test trench, the Douara Cave deposits and the other comparative deposits.

Samples

1: Fan gravels I and II, WF-2, 2: Fan gravels III and IV, WF-2, 3: Second Horizon, Douara Cave, 4: Unit A of Third Horizon, Douara Cave, 5: Unit B of Third Horizon and Units A and B of Fourth Horizon, Douara Cave.

Plotting area

A: Fan gravels I and II, B: Fan gravels III and IV, C: Second Horizon, D: Unit B of Third Horizon and Units A and B of Fourth Horizon, E: Deposits from sabkhet and nebkha zone in Palmyra Basin, F: Pebble beach deposits on Lebanese Coast.

of materials finer than −1 phi as matrix. On the other hand, Fan gravels III and IV of the second category include them in abundance as matrix, averaging 35% for Fan gravel III and about 50% for Fan gravel IV.

Proportions among the three components, gravel A (coarser than −4 phi, coarse pebble and cobble), gravel B (−4 to −1 phi, fine pebble and granule) and sand (finer than −1 phi, sand and coarse silt), are plotted on the ternary diagram of Fig. 4. 15 for samples from various deposits in and around the Douara Cave. Cave and alluvial fan deposits are scattered mainly in the inner portion of the diagram. It suggests that they are immature sediments, the cave deposits especially being the most unsorted among them, in contrast to the deposits in sabkhet, including the eolian deposits, or the beach deposits on the Lebanese Coast.

4. 5 SOME COMMENTS ON THE AGES OF THE DOUARA CAVE DEPOSITS

Radiocarbon datings of the cave deposits were carried out by Dr. W. G. Mook, Groningen University, and by Dr. Kobayashi, the University of Tokyo. The results are given in Table 4. 3, together with some stratigraphical remarks. Five samples of blackish hearth ash are taken from Unit B of the Fourth Horizon. Some lenses of blackish hearth ash from which the samples are collected, are rich in charcoal powder and interbedded in the whitish to brownish gray ash alternations. Sample TK-165 obtained from the uppermost horizon of the hearth ash alternation gave a younger age than those of the other four samples. The ages for samples TK-166 and TK-167 taken from the middle horizon of the hearth, are older than 43,200 y.B.P. TK-168 and GrN-7599 obtained from the lower horizon of the hearth are older than 43,200 or 52,000 y.B.P. It seems reasonable that the ages for the middle to lower horizon of the hearth are older than 43,200 or 52,000 y. B. P. But samples TK-165 from the uppermost horizon might have been affected by the contamination of younger materials, because there are no discrepancies in the cultural assemblages between these horizons. A sample of travertine taken from Unit B of the Third Horizon in the innermost part of the cave was dated by means of the fission track method (Suzuki, in preparation). A sample of ostrich eggshell fragments was collected from Unit B of the Third Horizon and has been prepared for radiocarbon dating.

Table 4.3. Radiocarbon Ages for the Deposits in the Douara Cave

Code No.	Dates	Material	Depth (cm)	Grid	Stratigraphical remarks
TK-165	38,900±1700	Hearth ash	110–120	9-04	uppermost IV B
TK-166	>43,200	Hearth ash	120–130	9-04	middle IV B
TK-167	>43,200	Hearth ash	130–140	9-05	middle IV B
TK-168	>43,200	Hearth ash	150–160	9-05	lower IV B
GrN-7599	>52,000	Hearth ash	140–150	8-03	lower IV B

4. 6 SOME DISCUSSIONS ON THE STRATIGRAPHY AND THE PALEOENVIRONMENT IN AND AROUND THE DOUARA CAVE

Fan deposits in the trenches of WF-1, 2 and 3 are similar and, as noted above, many of them are correlated with each other on the basis of the characteristics features of color, hardness, grain-size, calcareous crust, included flint implements and so on. Terrace gravel, the youngest sediment except for the Present gravel in the wadi floor, seems to have been deposited in the Pottery age. Fan gravel I of each trench consists of gravelly sediments which are well sorted, especially in WF-2. It was deposited before downcutting by Terrace gravel along the present wadi. Fan gravel II of each trench is a gravelly sediment in the upper and a comparatively massive sediment in the lower. Fan gravels I and II are considered to have been deposited in the Pre-Pottery Neolithic age. Fan gravel III of each trench is a massive and tight sediment pinkish to reddish-brown-colored, with frequent occurrences of calcareous crust. Included flint implements are purely of the Middle Paleolithic. There is an abundance of the flint implements of Levantine Mousterian, which are the same as those found in the Third Horizon of the Douara Cave. Almost all of them have worn-out edges and some are fragmental. They may have been derived from the Third Horizon of the Douara Cave. In addition, they are accompanied by a small number of flint implements which are patinated, dark colored, and fragmental. They resemble those found in the Fourth Horizon of the cave. Fan gravel III was deposited before the Epi-Paleolithic age and during or after the Middle Paleolithic age of Units A and B of the Third Horizon in the Douara Cave. Fan gravel IV of the trenches of WF-1 and 3 is a very tight and gravelly sediment cemented remarkably with calcareous crust. It was deposited in the same period as that of Fan gravel III. Fan gravels V and VI found only at WF-3 are very tight, reddish-brown sediments. As no flint implements were derived from the cave, they were probably deposited before the deposition of Unit B, the Fourth Horizon of the Douara Cave. This correlation between the cave deposits and the sediments around the cave is shown in Table 4. 4. Changes of the paleoenvironments in and around the Douara Cave are summarized as follows.

4. 6. 1 Before the Middle Paleolithic age

Fan gravels V and VI were formed. In the Cave, deposits of more than 3 m in thickness are considered to have formed. Some remnants of older alluvial fans distributed around the lower reaches from the cave might have been associated with these Fan gravels.

4. 6. 2 Middle to Upper Paleolithic age

During the Middle Paleolithic age, the Third and the Fourth Horizons formed in the cave. The Fourth Horizon is considered to have been deposited under cooler and less dry conditions than those of the present, as suggested by palynological results (Sohma and Sasaki, 1973) and the common occurrence of sedimentological features such as calcareous concretions, hearth layers and so on (Endo, 1973).

Prior to the Epi-Paleolithic age, probably during the latest Middle or the Upper

Table 4.4

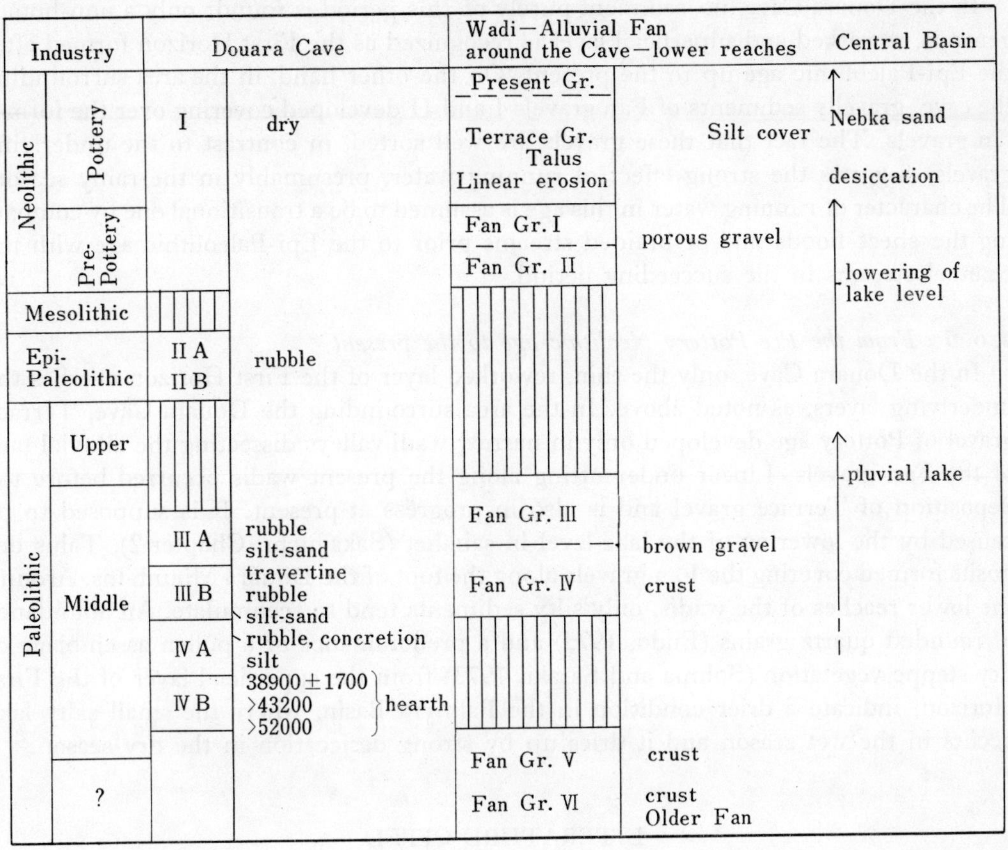

Paleolithic age, a great number of alluvial fans developed with the deposition of Fan gravels III and IV along the southern foot of the Douara Mountains. These Fan gravels, remarkably crusted but poorly sorted, are considered to have been deposited under humid conditions.

4. 6. 3 Epi-Paleolithic age

In the Douara Cave, gravelly deposits of the Second Horizon were laid down during this age, but around the cave, no sediment corresponding to the Second Horizon is found. According to palynological studies, the frequent occurrence of Chenopodiaceae and Artemisia, which are indicative of dry steppe vegetation, was recognized in the Second Horizon (Sohma and Sasaki, 1973). The rubble layer of the Second Horizon is very loose and fragmental and has only a small amount of sand and silt fraction. Moreover it contains a great number of rounded quartz grains as an indicator of drier conditions (Endo, 1973). These sedimentological characters of the Secont Horizon, which contrast remarkably with those of the underlying layer, suggest a drier environment than that prior to this age. It might have been much drier in summer, which is similar to the present conditions.

4. 6. 4 Pre-Pottery Neolithic age

In the Douara Cave, no sediment purely of this period is found; only a non-homogeneous, reworked and superficial layer is recognized as the First Horizon formed after the Epi-Paleolithic age up to the present. On the other hand, in the area surrounding the cave, gravelly sediments of Fan gravels I and II developed covering over the former fan gravels. The fact that these gravels are well sorted, in contrast to the underlying gravels, suggests the strong effect of running water, presumably in the rainy season. The character of running water in this age is assumed to be a transitional one by comparing the sheet floods and/or braided streams prior to the Epi-Paleolithic age with the channel streams in the succeeding period.

4. 6. 5 From the Pre-Pottery Neolithic age to the present

In the Douara Cave, only the thin, reworked layer of the First Horizon overlies the underlying layers, as noted above. In the area surrounding the Douara Cave, Terrace gravel of Pottery age developed only in narrow wadi valleys dissecting the alluvial fans of the Fan gravels. Linear undercutting along the present wadis occurred before the deposition of Terrace gravel and is also in progress at present. It is supposed to be caused by the lowering of the lake level in sabkhet (Sakaguchi, Chapter 2). Talus deposits formed covering the Fan gravels along the foot of the Douara Mountains. Around the lower reaches of the wadis, only silty sediments tend to accumulate. An abundance of rounded quartz grains (Endo, 1973) and a predominance of a pollen assemblage of dry steppe vegetation (Sohma and Sasaki, 1973) from the superficial layer of the First Horizon, indicate a drier condition in the Palmyra Basin, where the small salty lake occurs in the wet season and it dries up by strong desiccation in the dry season.

LITERATURE CITED

Akazawa, T., Baba, H., and Endo, K. (1973) Investigation of the Douara Cave Site, 1970 Season. The Palaeolithic Site at Douara Cave in Syria, Chap. II, 9–27, The University of Tokyo.

Endo, K. (1973) Sedimentological Analysis of the Deposits in the Douara Cave Site. The Palaeolithic Site at Douara Cave in Syria, Chap. IV, 98–111, The Universitiy of Tokyo.

Farrand, W. R. (1965) Geology, Climate and Chronology of Yabrud Rock Shelter I. Annales Archéologiques de Syrie, Vol. 15, Tome 1, 35–50.

Farrand, W. R. (1972) Geological correlation of prehistoric sites in the Levant. The Origin of Homo Sapiens (F. Bordes, ed.), 227–235, UNESCO, Paris.

Goldberg, P. (1969) Sediment Analysis of Jerf' Ajla and Yabroud Rock Shelters, Syria. Résumé de Communications, 8th INQUA Congress, Paris, 1969.

Matsutani, A. (1973) Microscopic Study on the "Amorphous Silica" in Sediments from the Douara Cave. The Palaeolitihic Site at Douara Cave in Syria, Chap. VIII, 127–132, The University of Tokyo.

Ponikarov, V. P., et al. (1966) The Geological Map of Syria and Its Explanatory Note. Ministry of Industry, Syrian Arab Republic.

Rust, A. (1950) Die Höhlenfunde von Jabrud (Syrien), Neumünster, 154 pp.

Schroeder, B. (1969) The Lithic Industries from Jerf Ajla and their Bearing on the Problem of a Middle to Upper Palaeolithic Transition. Doctoral Dissertation, Columbia University.

Sohma, K. and Sasaki, M. (1973) Palynological Studies on the Deposits of the Douara Cave. The Palaeolithic Site at Douara Cave in Syria, Chap. IX, 133–141, The University of Tokyo.

CHAPTER 5

Excavation at the Douara Cave

Banri Endo,* Tsuyoshi Fujimoto,** Takeru Akazawa***
and Kunihiko Endo****

* *Department of Anthropology, Faculty of Science, The University of Tokyo;*
** *Tokoro Field Laboratory, Faculty of Letters, The University of Tokyo;*
*** *Department of Anthropology, National Science Museum;* **** *Department of Applied Earth Sciences, Faculty of Sciences and Humanities, Nihon University*

5. 1 GENERAL REMARKS

The excavation of the second season was carried out at the Paleolithic site of Mugharat ad Douara I in 1974. This site, simply called the Douara Cave, is the biggest of some dozen caves located along the margin of the southern rocky slope of the mountain, Jabal ad Douara (or Dawwara). It is 18 km northeast of Palmyra, which is the biggest oasis in the Syrian Desert. The cave opens to the southeast, having a slope in front of it (Fig. 5. 1; Photo 5. 1a). For further details about the location, form, and surroundings of the site, as well as the findings of the excavation in the first season, see the previous reports of the Tokyo University Scientific Expedition to Western Asia (Suzuki and Kobori, 1970; Suzuki and Takai, 1973; 1974).

In this season, the main trench excavated in 1970 was widened in order to make further analysis of the Paleolithic industries and the nature of the deposits in relation to the contemporary circumstances and climates. Three test trenches were also excavated in front of the cave on the slope of the mountain, along a small wadi running out from the vicinity of the cave. The purpose of the excavation outside the cave was to investigate the relationship between the deposits in the cave and those of the slope of the mountain. The spatial relation of these excavated localities is shown in Fig. 5. 1.

The excavation work in the Douara Cave was directed by T. Akazawa on the prehistoric aspect, while the work on the part of the deposits yielding the Upper Paleolithic industry was directed by T. Fujimoto. At the same time, the deposits were examined by K. Endo from the sedimentological point of view. The excavation of the test trench outside the cave for the sedimentological investigation was directed by K. Endo. The engineering work to meet the requirements of prehistoric and sedimentological investigations, such as the setting of the coordinate system for the various measurements needed in the excavation and making of the lacquer-film preparation of the stratigraphy, was done by B. Endo.

A minute method of excavation, such as was done by Tixier (1970), might have been the best way to obtain precise data from the excavation, but we had only limited time and finances. Thus, for this season, we selected the method of excavating unit-block by unit-block.

a

b

Photo 5. 1. a. Distant view of the Douara Cave.
b. Grid for excavation set in the Douara Cave.

5. 2 METHOD OF EXCAVATION IN DOUARA CAVE

A datum point is always necessary for measurement in an excavation. The point was already marked in 1970 on the surface of the wall at the center of the back of the cave. It was designated as Point (7,0,0) in [m] in the three dimensional XYZ-coordinate system, in which X and Y were horizontal axes crossing each other at right angles, whereas Z was the perpendicular axis. The Y axis was made parallel to the longitudinal axis of the cave. The relationship between the coordinate and the cave is shown in Fig. 5. 2. The datum point was 552.8 m above sea level, and the horizontal axis of Y pointed to E 19°S. The same coordinate had already been used in 1970.

A new hole, with a diameter of 6 mm, was made with a stone-drill in the wall at the datum point and a small bolt was planted in the hole with a lead attachment. An iron post was perpendicularly erected with cement on the rock at a point along a YZ plane, passing through the datum point. Then, a horizontal grid, representing the XY coordi-

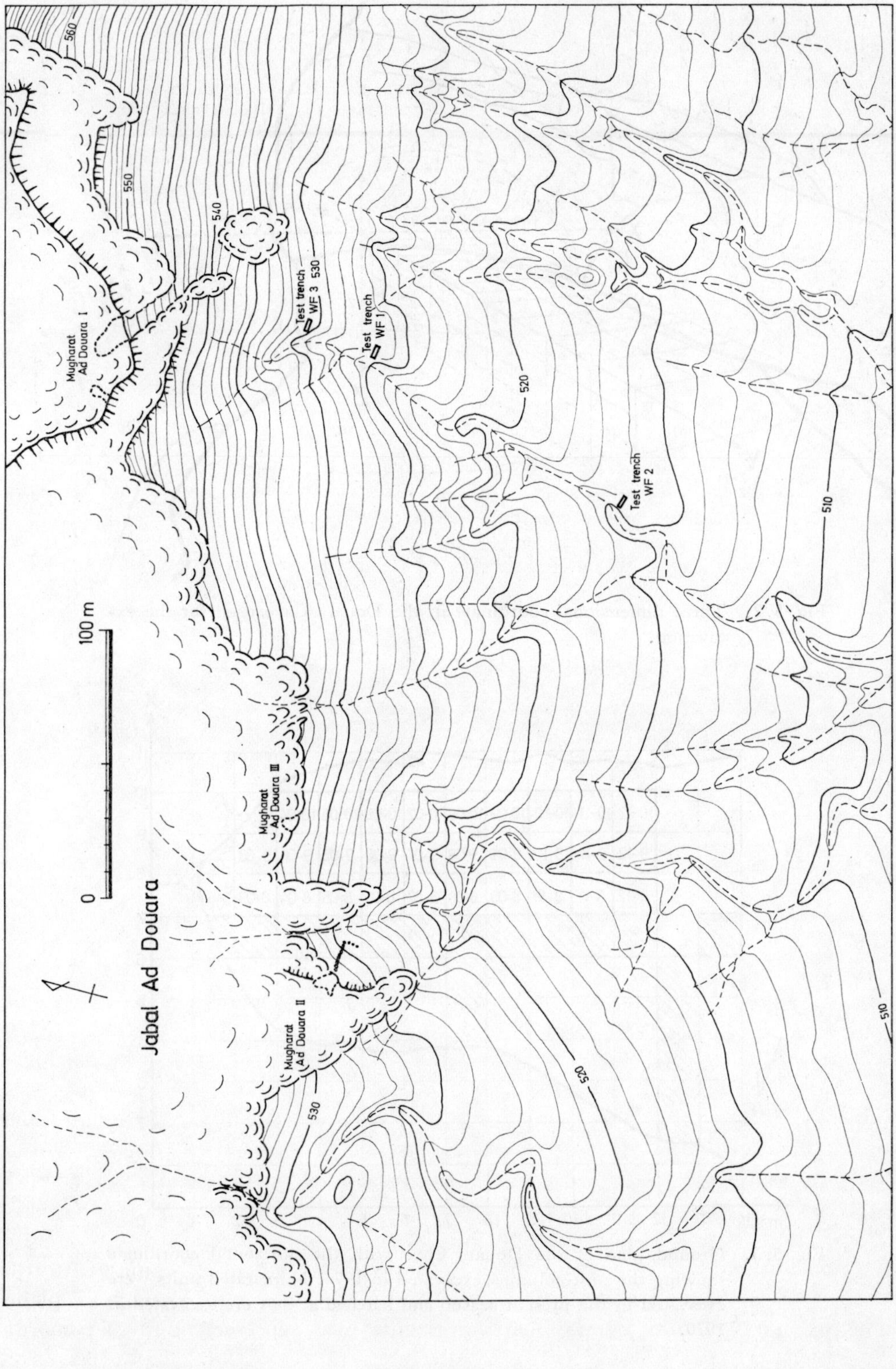

Fig. 5. 1. Map of the southern foot of Jabal ad Douara and the Douara Cave.

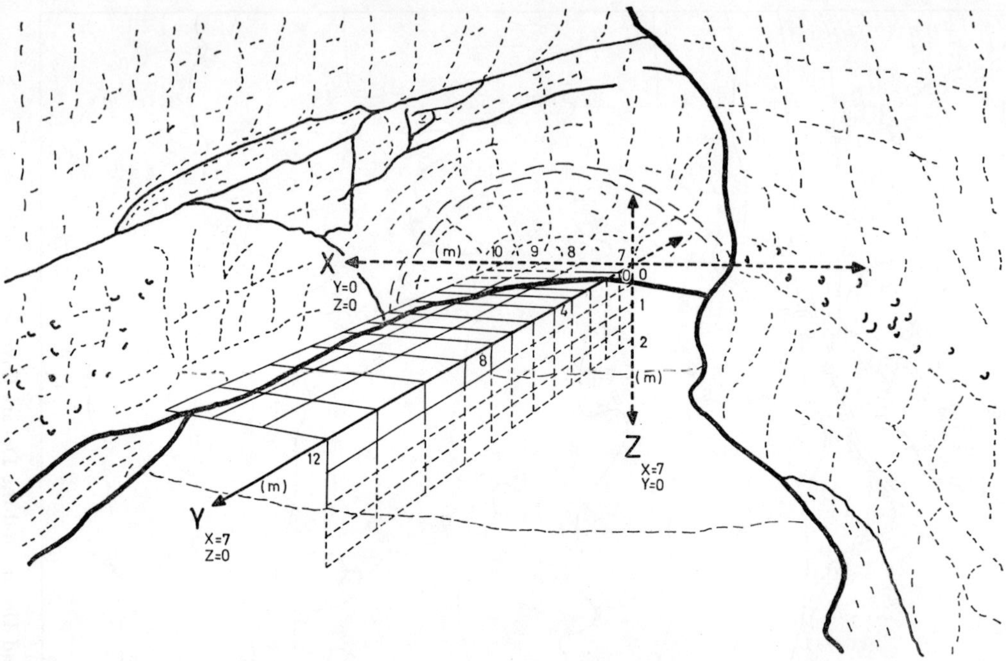

Fig. 5.2. Three dimensional coordinate in the Douara Cave used for the excavation.

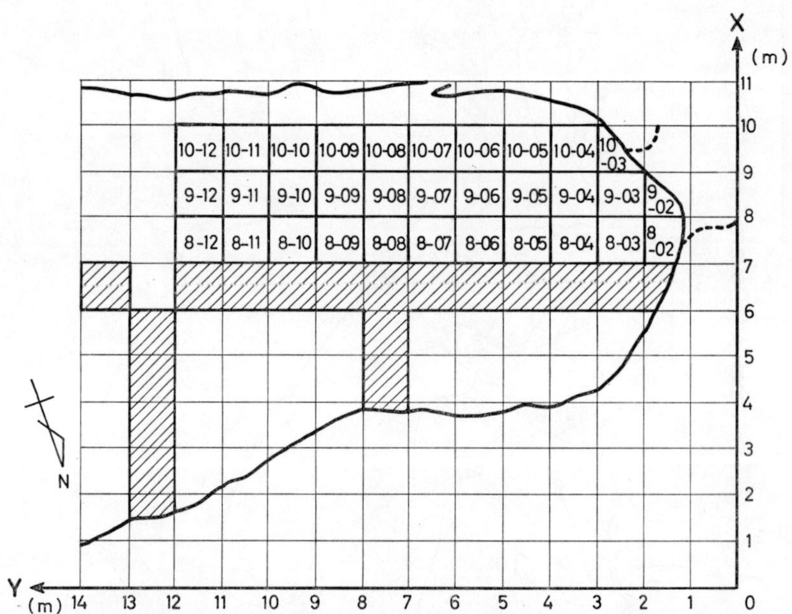

Fig. 5.3. Ground plan of the Douara Cave with the horizontal coordinate showing the unit-columns excavated in 1974. Numerated units were excavated in the present season and hatched areas were excavated in 1970.

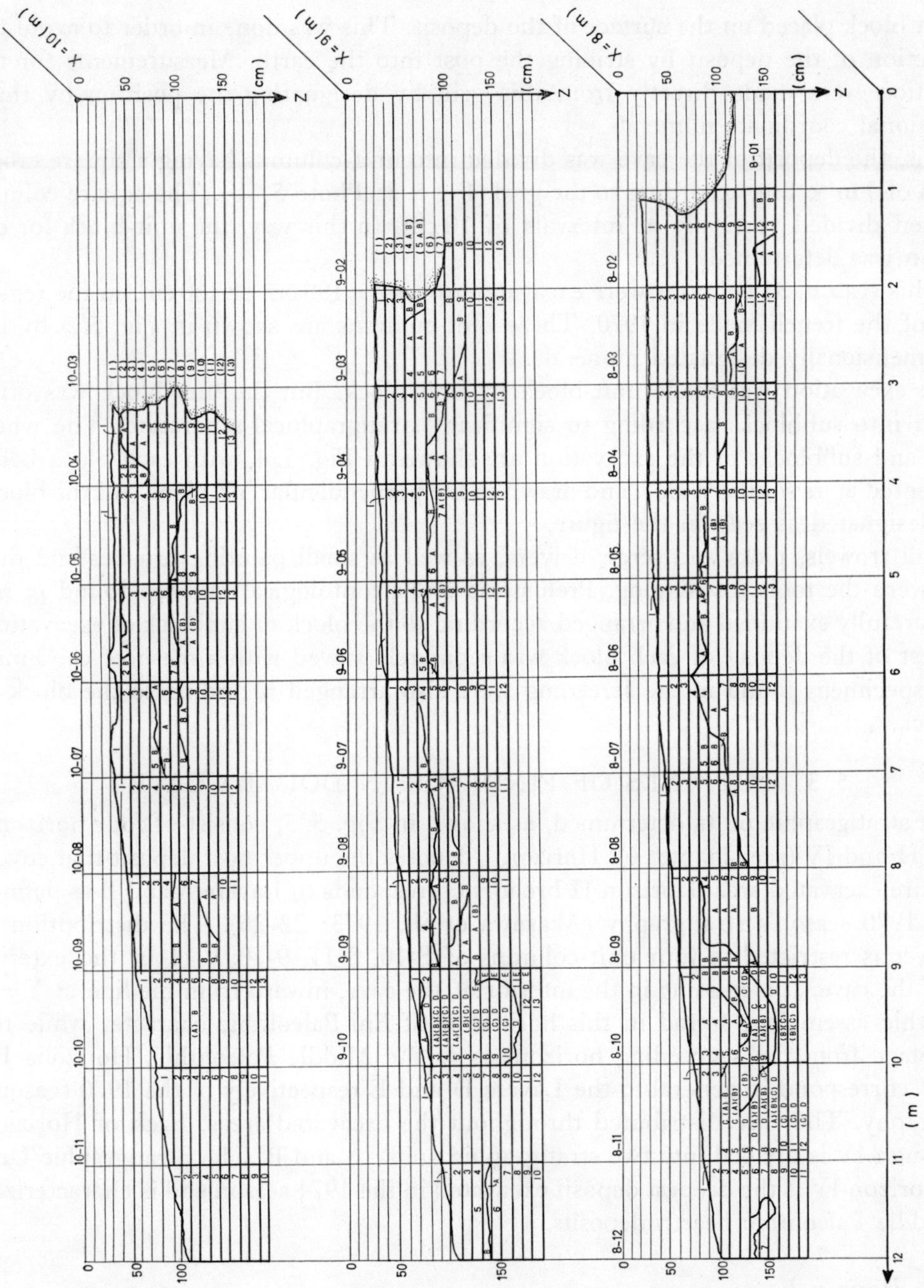

Fig. 5. 4. Scheme of the unit-blocks and subblocks shown on the perpendicular YZ plane. Letters in parenthesis are unvisible on the side of the block shown in the figure.

nate, was made at an interval of 1 m. The actual grid consisted of wooden beams and rubber strings, fixed either to the bolt embedded into the wall of the cave or to the wooden post. The wooden beam was suspended by wires to keep it straight in the same way as a suspension bridge. The bottom of the post was embedded in the heavy square cement block placed on the surface of the deposit. This was done in order to avoid the destruction of the deposit by striking the post into the earth. Measurements for the excavation were made directly from this grid by designating the position by three dimensional coordinate numerals.

Thus, the deposit of the cave was divided into unit-columns having a square cross-section of 1 m × 1 m, according to the grid (Fig 5. 3; Photo 5. 1b). The square column was then divided vertically at intervals of 10 cm. In this way, the unit-block for excavation was determined.

In this season, 33 columns were excavated up to Z=190 cm or 160 cm, in the region south of the trench made in 1970. These unit-columns are shown in Fig. 5. 3 by the two-dimensionally designated numerals.

The excavation was made unit-block by unit-block. But the unit-block was often divided into subblocks according to significant stratigraphical separation. The whole blocks and subblocks of the excavation are shown in Fig. 5. 4, with each block being represented at its southern side, and drawn in the perpendicular YZ plane. All the blocks were designated as seen in the figure.

Small trowels, picks and screw-drivers, as well as small painting brushes and dust pans, were the tools for digging. Prehistoric or Paleontological findings found *in situ* were carefully examined and arranged according to the block or subblock of excavation. The rest of the deposit of each block was separately sieved with a mesh of ca. 3 mm. Small specimens found in the screening were also arranged according to the block or subblock.

5. 3 OUTLINES OF FINDINGS IN DOUARA CAVE

The stratigraphic units determined, as shown in Fig. 5. 5, consist of four horizons: I, II, III and IV (see Chapter 4). Horizon I denotes the uppermost deposit that covers the entire excavated area. Horizon II broadly corresponds to Layers A and B as defined in the 1970 season's stratigraphy (Akazawa et al., 1973: 22–26). The distribution of this layer is restricted to four unit-columns of 8-10, 8-11, 9-10, 9-11 in the exterior part of the cave. It is missing in the interior of the cave, inward from the line at Y=9. The lithic assemblage found in this horizon is of Epi-Paleolithic character while the assemblage from the succeeding horizons is of the Middle Paleolithic. Horizons III and IV correspond in general to the Layers D and E respectively of the 1970 season's stratigraphy. They are distributed throughout the excavated areas. Each of Horizons II through IV is divided into two stratigraphic units: A and B. The Stratigraphic Unit B of Horizon IV is the deepest deposit excavated in the 1974 season and is characterized by Middle Paleolithic hearth deposits.

5. 3. 1 Horizon I

The collection of artifacts from this horizon is not homogeneous. Although it is a very thin deposit, it contains a small number of Epi-Paleolithic and Neolithic artifacts such as naviform type cores, as well as a large quantity of Middle Paleolithic artifacts. In addition, it contains some potsherds, glass, and bronze tools that presumably belong to the Roman and Islamic periods. The above lithic assemblages of Middle and Epi-Paleolithic types have the same morphological and technological features as those of

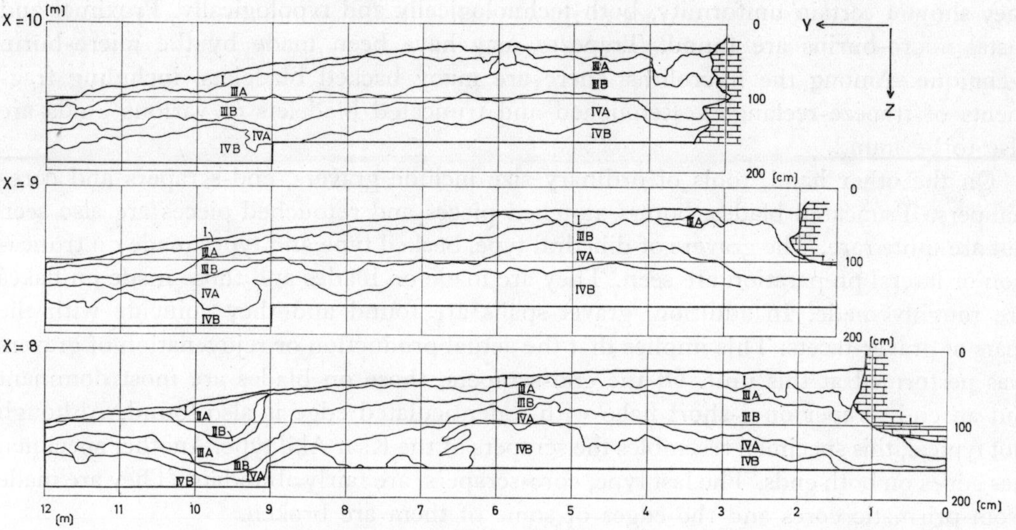

Fig. 5. 5. Scheme of the stratigraphy of the deposit in the Douara Cave.

assemblages from the succeeding horizons. But no other horizon yields the lithic artifacts of Neolithic types or potsherds and bronze tools.

The physical condition of flints from this horizon is markedly different from that of other horizons. Included in the lithic collection are a large quantity of small fragmental pieces probably broken as a result of thermal action, but which are definitely not chips. Almost all of the flints are more or less abraded and usually patinated black to very dark brown with a luster. On the whole, these contrast with the fine, clean condition and usually brown color of the flints from the underlying Horizon III. Such context and condition of Horizon I artifacts imply that this horizon may not be a primary deposit.

5. 3. 2 *Stratigraphic Unit A of Horizon II (Unit II A)*

The collection from this unit is dated to the Epi-Paleolithic. The artifacts here have more calcareous concretion than those of Unit B and they are light brown to black in color and have a white cortex. The total collection consists of retouched pieces and blanks of various kinds. Blanks include flakes, blades, bladelets, cores, and core-rejuvenated flakes, and they are highly microlithic.

Flakes occupy four fifths of the total quantity of artifacts. Blades and bladelets are mostly broken and are classified as parallel-sided blanks made by punch-technique: those less than 15 mm in widths become bladelets and the rest are blades. However, this classification is arbitrary.

Cores are found in fairly large numbers. They are made on large flakes or flint pebbles and mostly are of the prismatic type. Many are uni-directionally flaked with single-faceted platforms. Fairly large numbers of core-rejuvenated flakes are also found in association.

Retouched pieces are very few in number, occupying only 2% of the total quantity of artifacts, and of these, two thirds are microliths. Among the microliths, trapeze-rectangles are characteristic and are the dominant type. Most of them are broken but they show a certain uniformity, both technologically and typologically. Proximal and distal micro-burins are found. Trapezes may have been made by the micro-burin technique. Among the microliths, there are many backed bladelets, including fragments of trapeze-rectangles. Retouched and truncated bladelets of various kinds are also to be found.

On the other hand, tools of ordinary size include gravers, end-scrapers and core-scrapers. Truncated blades, borers, notched pieces and retouched pieces are also seen but are quite rare. The gravers of dihedral type, beaked type and types made on truncation or lateral preparation are seen. They are made on blades and those made on flakes are roughly made. In addition, graver-spalls are found and they coincide with the scars of graver-facets. This implies that the actual production or rejuvenation of gravers was performed at this unit. Of the end-scrapers, those on blades are most dominant but an end-scraper on a short flake with a denticulated edge is also found. Although not typical, this specimen resembles the scrapers of the Ksar Akil type. Another specimen has edges on both ends. The last type, core-scrapers, are fairly abundant. They are made from prismatic cores and the edges of some of them are broken.

5. 3. 3 Stratigraphic Unit B of Horizon II (Unit II B)

Specimens collected from this unit are datable to the Epi-Paleolithic. Artifacts in this unit are almost identical to those of Unit II A. They are light brown to black in color, with a white cortex, and have light calcareous concretion on their surfaces. Many artifacts, especially cores and large tools, have a cortex.

Flakes are most abundant and account for seven eighths of the total quantity of specimens. The next most common specimens are bladelets. Blanks in this collection are highly microlithic.

Cores and core-rejuvenated flakes are present, too. Cores are of the prismatic type, and most of them are uni-directionally flaked. Most of them have been modified into core-scrapers.

Of the retouched pieces, microliths are dominant and characteristic. They number less than 2% of the total quantity of artifacts. Among the microliths, trapeze-rectangles are the most characteristic tools. They are similar to one found in Unit A. As in that unit, trapezes may have been made by the micro-burin technique in this unit. In addition, there are a large number of backed bladelets in broken pieces. On them, there are various types of retouch. Proximal and distal micro-burins are seen, too. However, they may be merely waste as a result of trapeze production.

As in Unit A, various gravers are seen. They are dihedral or beaked or are made on broken end or lateral preparation. The end-scraper is atypical and is made on a flake. It has edges at both ends: at the proximal end on the ventral surface and at the distal end on the dorsal surface. In association, core-scrapers are found in fair number. They are modified from prismatic cores. No borers, truncated pieces, or typical end-scrapers are seen in this unit.

5. 3. 4 Epi-Paleolithic assemblage

Some differences are observed on several points between the assemblages of Unit II A and Unit II B, but these two assemblages are basically the same. The ratio of blanks is similar and the assemblages of both units have a large number of non-retouched pieces, these being one of the most characteristic features of the assemblages. The ratio between blades and bladelets, whose production systems are precisely the same, is identical. They are flaked from uni-directional and prismatic cores. Thus, technologically, the assemblages of both units are identical.

The blanks and retouched pieces of Units II A and II B are microlithic. Their most characteristic tools are trapeze-rectangles. They have a certain quantity of micro-burins. It is probable that trapezes of these units are formed by the micro-burin technique. There is a difference in the ratio of microliths, but the same types are found in both assemblages.

In ordinary-sized tools, the assemblages of both units have the same types of gravers. Dihedral and beaked gravers and types of gravers on lateral preparation are seen in both assemblages. The ratios of types of gravers are similar. Core-scrapers, modified from the prismatic type of cores, are seen in both assemblages. Retouched and notched pieces are also seen in both assemblages.

The differences between these two assemblages are observed in borers, truncated pieces, and typical end-scrapers on blades. These are not found in the assemblage of Unit B. Likewise, a scarcity of scrapers is observed in the assemblage of Unit B. The assemblages of both units, especially Unit B, have small number of tools. It is probable that the differences between the assemblages of Units A and B have arisen merely accidentally. In the investigation of the 1970 season, borers and truncated pieces were found in Layer B. Thus, the two assemblages also coincide from the typological point of view.

As for traces of wear, the same traces are observed at the same positions in the same type of tool at both assemblages.

As mentioned above, concerning the technological, typological and functional features of tools, the identity is observed between the assemblages of the Units A and B. It is summarized as follows:

1) A large amount of non-retouched pieces are seen, and cores and core-rejuvenated flakes are found in fairly large numbers.

2) The assemblages of both units are microlithic in blanks and tools. In retouched tools, microliths occupy about two thirds of the total. Among microliths, trapeze-rectangles with widths of 6–9 mm are abundant. Trapezes may have been made by the micro-burin technique.

3) The ordinary-sized tools occupy one third of the total. Among ordinary-sized tools, gravers of various types, end-scrapers on blades and core-scrapers are representative.

The assemblages of Unit II A and Unit II B have the features mentioned above. They coincide with the general features of Geometric Kebaran A of Bar-Yosef (Bar-Yosef, 1970; 1975; 1976), especially in the aspects of having fairly large numbers of trapeze-rectangles. A detailed description of these assemblages will be given in a final report.

5. 3. 5 Stratigraphic Unit A of Horizon III (*Unit III A*)

The artifacts found in this unit are not as homogeneous as those of the underlying units. The majority of the lithic artifacts closely resemble those of the Levantine Mousterian assemblage from the unit below. But mixed with them are very few potsherds and bronze tools and flint implements of non-Paleolithic character. It is also noteworthy that a considerable proportion of flints are broken and have a black to very dark brown patina, presumably caused by fire.

This evidence suggests the possibility of this unit being a secondary or disturbed deposit. However, this question requires further analysis of the geology in order to determine the exact nature of this unit, including its relationship to the associated lithic assemblage.

5. 3. 6 Stratigraphic Unit B of Horizon III (Unit III B)

This unit produced the largest quantity of lithic artifacts of all the stratigraphic units of Horizons III and IV. The prominent feature of the assemblage of this unit is the high frequency of the Levallois technique used in the production of tool blanks and the high percentage of blanks having faceted striking platforms pertaining to the Levantine Mousterian industries as defined in a number of sites in Western Asia. The physical condition of these flints is usually clean, unabraded and of bright color in contrast to lustrous and dark color of flints from the overlying deposits.

A large number of ostrich (*Struthio* sp.) eggshell fragments are found in the deposits. Although they are found in both Unit III A and Unit IV A as well, they are concentrated mostly in this unit. All in all, Unit III B is characterized by deposits of a rich Levantine Mousterian assemblage as well as a large number of ostrich eggshell fragments, without the associated non-Paleolithic flints and potsherds.

5. 3. 7 Stratigraphic Unit A of Horizon IV (Unit IV A)

From this unit, the lithic artifacts decreased in number. The lithic assemblage is broadly characterized as exhibiting Levantine Mousterian features. But on closer examination, its features are found to be different from those of the overlying Unit III B. One of the most striking features of this assemblage is the relatively low frequency of the Levallois technique in the production of tool blanks. This contrasts with the dominance of the Levallois technique in the manufacture of blanks in Unit III B. The above tendency is observed more clearly in the assemblage of the succeeding Unit IV B.

The abrasion and patination of the flints of this unit are variable. The outstanding physical characteristic of the flints is that a considerable number of flints show some abrasion and are in a brittle condition, presumably caused by fire, and are patinated bluish to very dark and blackish-gray in color. This evidence is also emphasized in the following assemblages from Unit IV B.

5. 3. 8 Stratigraphic Unit B of Horizon IV (Unit IV B)

This unit corresponds to the hearth deposits containing a number of ash beds and hearth streaks in the matrix of whitish or grayish to blackish stained earth mixed with some organic matter. The hearth band is very thick, ranging from 50–60 cm in the thickest parts (8-04 to 8-06, 9-04 to 9-06) to 20-30 cm and is outlined by an arrangement of limestone fragments and a series of very compact concretions. The stones are usually reddish yellow and are brittle as a result of thermal action. A dense ash concre-

a

b

Photo 5. 2. a. Big hearth found in the Douara Cave.
b. Remained ash in the hearth.

tion is found in the area defined clearly by a series of these stones and concretions (Fig. 5. 6; Photo 5. 2a, b).

The lithic assemblage from this unit is markedly different from that of the overlying Unit III B. The flints are usually patinated bluish or blackish-gray, or very dark gray. A large number of them show breakage, perhaps caused by fire, and are easily broken. The frequency of the Levallois technique drops. A large proportion of these blanks are characterized by an elongated form with a non-faceted striking platform. This contrasts with the significant presence of multidirectional, broad flakes of the Levallois type in the manufacture of blanks in Unit III B.

5. 3. 9 *Middle Paleolithic assemblage*

Finally, the relationship between the Middle Paleolithic assemblages of the two horizons, III and IV, can be broadly summarized as follows:

1) The figure of the Levallois Index of Horizon III is significantly larger than that

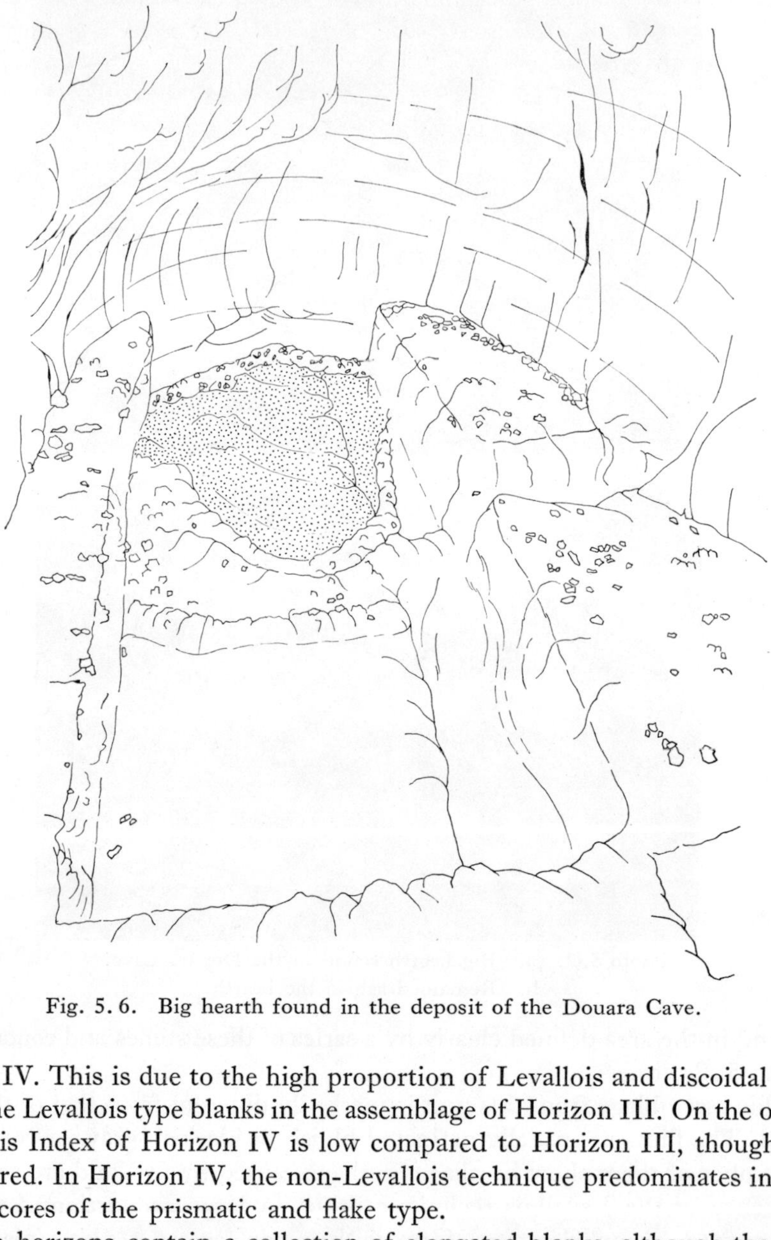

Fig. 5. 6. Big hearth found in the deposit of the Douara Cave.

of Horizon IV. This is due to the high proportion of Levallois and discoidal type cores as well as the Levallois type blanks in the assemblage of Horizon III. On the other hand, the Levallois Index of Horizon IV is low compared to Horizon III, though it should not be ignored. In Horizon IV, the non-Levallois technique predominates in the forms of specific cores of the prismatic and flake type.

2) Both horizons contain a collection of elongated blanks, although the quantities differ. Horizon IV holds an exceedingly high Blade Index, a tendency which is unlikely in the Middle Paleolithic context of the Levant. Of the elongated blanks, the majority are produced by the non-Levallois technique in Horizon IV and the Levallois technique in Horizon III.

3) Secondarily retouched tools are rare in both horizons. Since there are very few Levallois type blanks in Horizon IV, they naturally have very few unretouched tools, that is Levallois flakes and points. This evidence makes it clear that the assemblages

of Horizon IV are typologically dissimilar to any known Levantine Mousterian industry. In contrast, the typological characteristics of Horizon III readily conform to the Levantine Mousterian assemblages.

5.4 EXCAVATION AT WADI FAN

The excavation followed a grid system. Each of the trenches was made into grids of one meter square; each unit was excavated in 10 through 50 cm arbitrary levels, and all artifacts within these units and levels were recorded.

Thus, during 1974 the excavation was carried out in three trenches designated as WF-1, WF-2, and WF-3. WF-1 was composed of six units, WF-2 of five units and WF-3 of seven units. The depth of the excavations varies from 0.25 m to about 3 m (Figs. 5. 7/9).

In these trenches a total area of 18 m² was excavated. About 35 m³ of the deposits was removed in all. Excavated sediments were screened through a mesh of ca. 3 mm to increase atrifacts recovery. Among very large artifacts, numerous small-sized artifacts were recovered.

As shown in Fig. 5. 10, the stratigraphic units of the WF trenches consist of six layers: Fan gravels I through VI (see Chapter 4). With the exception of a very small collection from Fan gravels V and VI, the material of each layer is in sufficient quantity for us to compare those assemblages quantitatively and qualitatively within themselves and with other assemblages from relevant sites such as the Douara Cave. The general features of the collections from each layer are as follows.

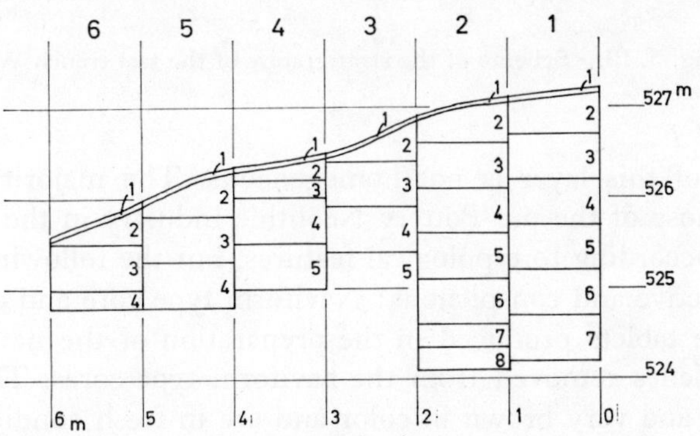

Fig. 5. 7. Blocks of excavation of the test trench WF-1.

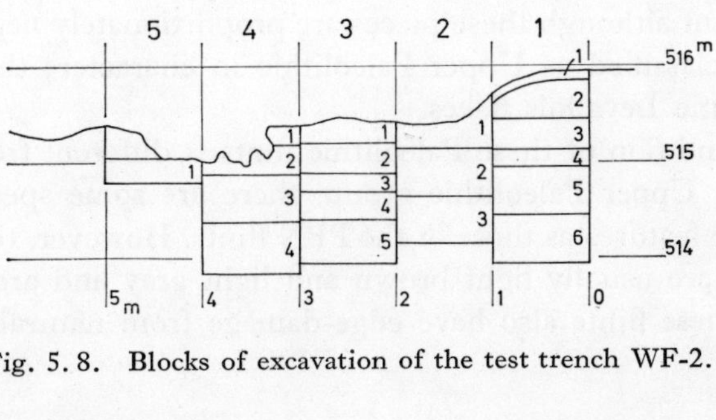

Fig. 5. 8. Blocks of excavation of the test trench WF-2.

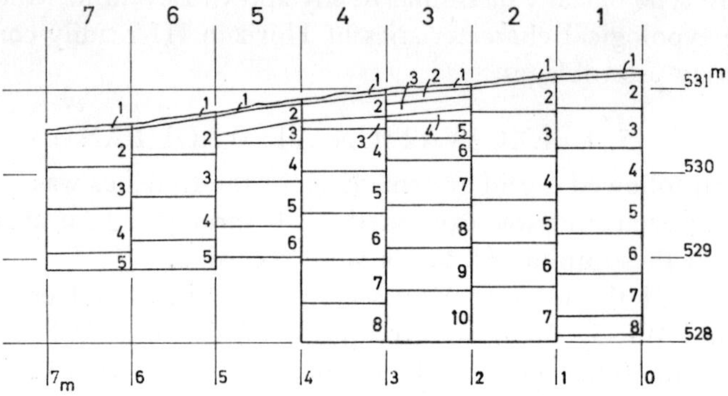

Fig. 5.9. Blocks of excavation of the test trench WF-3.

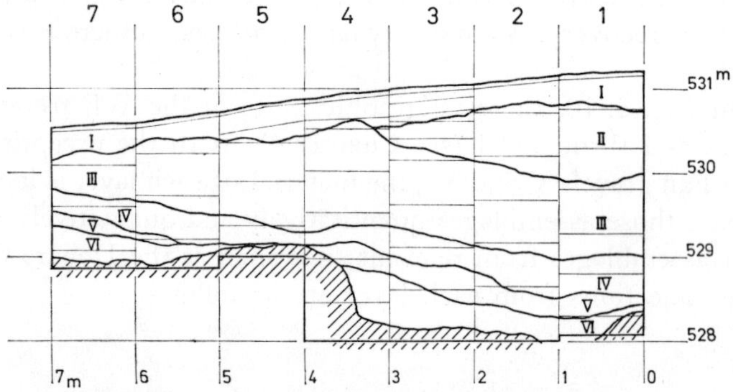

Fig. 5.10. Scheme of the stratigraphy of the test trench WF-3.

5.4.1 *Fan gravel I*

The collection of this layer is not homogeneous. The majority of lithic artifacts closely resemble those of the pre-Pottery Neolithic industry in the Levant. These fall into several types according to typological features, but the following types of artifacts are the most distinctive and conspicuous: Naviform type core and crest blades, plunging flakes and core tablets produced in the preparation of the naviform cores, and a large number of blades removed from the naviform type cores. These PPN flints are mottled mahogany and very brown in color and are in fresh condition, but almost all pieces have extensive edge-damage along their margins from natural crushing.

Together with these, a very small number of Upper and Middle Paleolithic flakes and cores are present although these pieces are proportionately negligible. Some blades and cores can be classified as Upper Paleolithic in character; the Middle Paleolithic flints consist of some Levallois flakes.

The physical condition of these Paleolithic flints is different from that of the above PPN flints. In the Upper Paleolithic group, there are some specimens characterized by having the same features as those of the PPN flints. However, the majority of Upper Paleolithic groups are usually light brown and light gray and are more abraded than the PPN flints. These flints also have edge-damage from natural crushing.

5. 4. 2 *Fan gravel II*

Again, in this layer, PPN type flints predominate in the total collection. The techno-typological features and the physical condition of these types of flints are exactly the same as those of the overlying layer.

The outstanding feature of the collection in this layer is that the proportion of the Middle Paleolithic type of flint is greater than that of the preceding layer. Among the total samples identified clearly, more than 20–30% consist of Levallois cores and flakes of all types. The physical condition of these Levallois type flints is light to brown and dull brown in color, in contrast to the flints of the associated PPN type with mahogany and dark brown color.

The next striking feature of the collection in this layer is that a small number of bluish and very dark colored flints are to be found. Although these flints of peculiar color are all fragmentary, all of them are classifiable as Middle Paleolithic type flints. In general, the materials are abraded and rolled, and all of them have small patches of calcareous concretions adhering to their surfaces. The features of these flints closely resemble the bluish to blackish and very dark gray flints of Levallois and non-Levallois type associated with Unit B of Horizon IV of the Douara cave. Their appearance suggests that they are derived from the Douara cave.

It is clear that Upper Paleolithic type flints are included in this collection, but the proportion of these types is completely negligible.

5. 4. 3 *Fan gravel III*

The nature of the collection from this layer is completely different from that of the preceding layers. The most striking feature is that the whole collection consists of various Levallois cores of Middle Paleolithic type and various Levallois flakes. Specimens classifiable as Levallois elements are mostly typical. Many of the Levallois flakes are of the broad type and are generally characterized by multidirectional and elaborate preparation of the striking platforms. The outstanding feature of these Levallois elements is their patination, which is usually a dull white and cream color, and the extensive edge-damage along their margins from natural crushing. These Levallois flints of all types closely resemble the Levallois cores and flakes discovered in Unit B of Horizon III of the Douara cave, with the exceptional difference of their physical condition.

Again, in this layer, fragments of bluish to very dark colored flint are included. Their features are exactly the same as those of the preceding Fan Gravel II, but the number of these fragments increases in this layer. At the top part of this layer, in particular, they are found in large number and show a slight decrease going downwards.

Lastly, there is a great number of small fragments found in this layer. In particular, thousands of these fragments of various colors and degrees of abrasion are collected from Fan Gravel III of the WF-1 trench. Their breakage is caused by natural crushing.

5. 4. 4 *Fan gravel IV*

The number of flints from this layer is significantly smaller compared to the overlying layers. All the specimens classifiable definitely fall into Middle Paleolithic of the Levallois type. The techno-typological characteristics of these Levallois type flints are broadly the same as those of the overlying Fan Gravel III. However, it is not possible to conclude whether the assemblages resemble one another because the sample size of Fan Gravel IV is small and almost all the pieces are fragmentary. All specimens are patinated in a deeper white to cream color than the flints from Fan Gravel III. Edge-damage of the flints is extensive.

Bluish and very dark-colored flint fragments mentioned in the preceding Fan Gravels II and III were not discovered in this layer, although occasionally such flints were collected, suggesting a mixture with the flints of the overlying layer presumably caused accidentally during excavation.

LITERATURE CITED

Akazawa, T., Baba, H. and Endo, K. (1973) Investigation of the Douara Cave Site. In: Suzuki and Takai (eds.), The Palaeolithic Site at Douara Cave in Syria. Part I. Univ. Mus., Univ. Tokyo, Bull. No. 5, pp. 9–27.

Bar-Yosef, O. (1970) Prehistoric Sites near Ashdad, Israel. Palestine Exploration Quarterly, 1970 Jan.–June, pp. 53–69.

Bar-Yosef, O. (1975) The Epipaleolithic in Palestine and Sinai. In: Wendorf and Marks (eds.), Problems in Prehistory: North Africa and the Levant. Dallas. pp. 363–378.

Bar-Yosef, O. (1976) A Note on the Geometric Kebaran A. Preconference papers for IXth International Congress of Prehistoric and Protohistoric Sciences, Nice, Colloque III, pp. 78–105.

Suzuki, H. and Kobori, I. (eds.) (1970) Report of the Reconnaissance Survey on Palaeolithic Sites in Lebanon and Syria. Univ. Mus., Univ. Tokyo, Bull. No. 1.

Suzuki, H. and Takai, F. (eds.) (1973) The Paleolithic Site at Douara Cave in Syria. Part I, Univ. Mus., Tokyo, Bull. No. 5.

Suzuki, H. and Takai, F. (eds.) (1974) The Palaeolithic Site at Douara Cave in Syria. Part II, Univ. Mus., Univ. Tokyo, Bull. No. 6.

Tixier, J. (1970) L'Abri sous Roche de Kasr ʿAquil. La Campagne de Fouilles 1969. Bull. Mus. Beyrouth, T. 23, pp. 173–191.

CHAPTER 6

The Assemblages of Site 50

Tsuyoshi Fujimoto
Tokoro Field Laboratory, Faculty of Letters, The University of Tokyo

This article is a preliminary report on the assemblages of Site 50. The analysis of artifacts is still going on. The detailed descriptions of the assemblages will be given in Part II of this report.

Site 50 lies on the northwestern terrace of Lake Sabkha (Sabkhet Mouh). On the edge of the terrace, there is the well named Ouinat Mohmed el Fard. This well is utilized for the growing of dates (Fig. 6. 1).

Surrounding the well are five spots in which concentrations of artifacts were found. They are named Spots A, B, C, D, and E respectively. Outside these concentrated areas, artifacts diminished remarkably. About 200 m northeast of the well, another spot in which artifacts were concentrated was found. It is named Spot F. The spots vary in diameter from 5 to 17 m. Site 50 was found in the geographical survey. The surface-collecting at this site was carried out in three days. At first, the area in which artifacts were concentrated was delimited and mapped. Then, the surface-collecting of artifacts began. Surface-collecting of artifacts was carefully carried out, because they included a large number of microliths. The flint from which the artifacts were made was brown to black in color. The surfaces of the artifacts were weathered or glazed, with some variation among the different spots. There were fairly distinct differences among the assemblages from each spot.

6.1 THE ASSEMBLAGE OF SPOT A

This is the spot surrounding the well (Fig. 6. 1). For about 14 m around the well, there were blocks of soil containing many artifacts. They were dug up in reconstructions of the well. The artifacts included in the blocks were also dug up. They are different from those of the other spots. In this spot, core-sampling of the soil from the wall of the well was also carried out. There were many artifacts in these cores.

Most of the artifacts in this spot have glaze on their surfaces. This glaze is very different from that of the other spots and may have been caused by water activity. The surfaces of some artifacts were so glazed that their ridges were dissolved away. In addition, some artifacts were abraded. It is concluded geographically that the deposits around the well may be not primary but secondary deposits. The artifacts in this spot have a certain uniformity from the techno-typological point of view. Even though they are not deposited primarily, they can be regarded as a single assemblage.

Artifacts (Fig. 6. 2): There are fairly large number of microliths among the retouched tools, but ordinary sized tools are superior in number. Among the microliths, finely backed bladelets and truncated bladelets are characteristic.

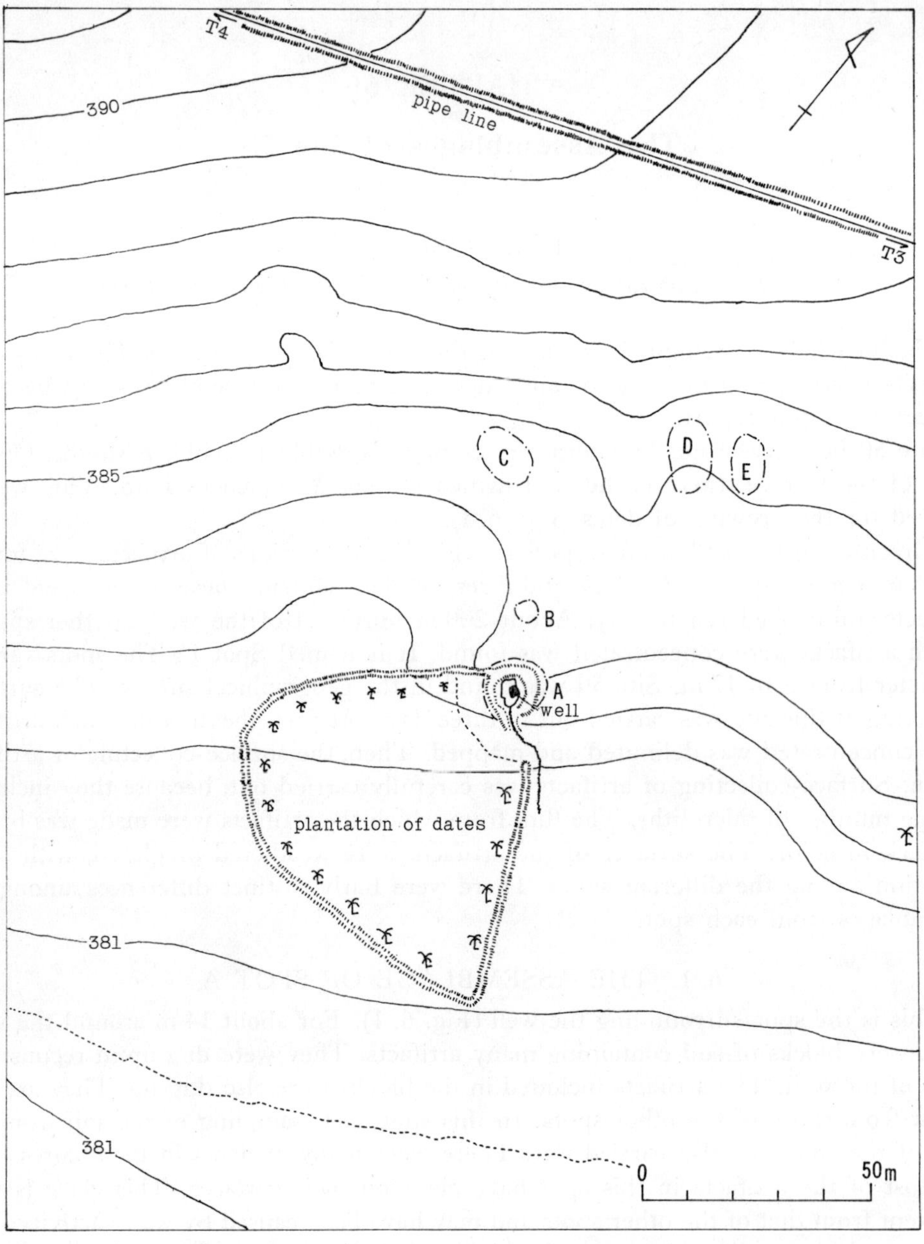

Fig. 6.1. The map of Site 50 (—·— artifacts concentraing areas, Spots A, B, C, D and E).

Finely backed bladelets are finely backed pieces on one side of bladelets (Fig. 6. 2. 1–11). The typical specimen has a width of 4–6 mm and is made on a slender type of bladelets.

Truncated bladelets are made on the bladelets actually parallel sided and about 10 mm in width (Fig. 6. 2. 12–17). They have perpendicular and straight truncations. The

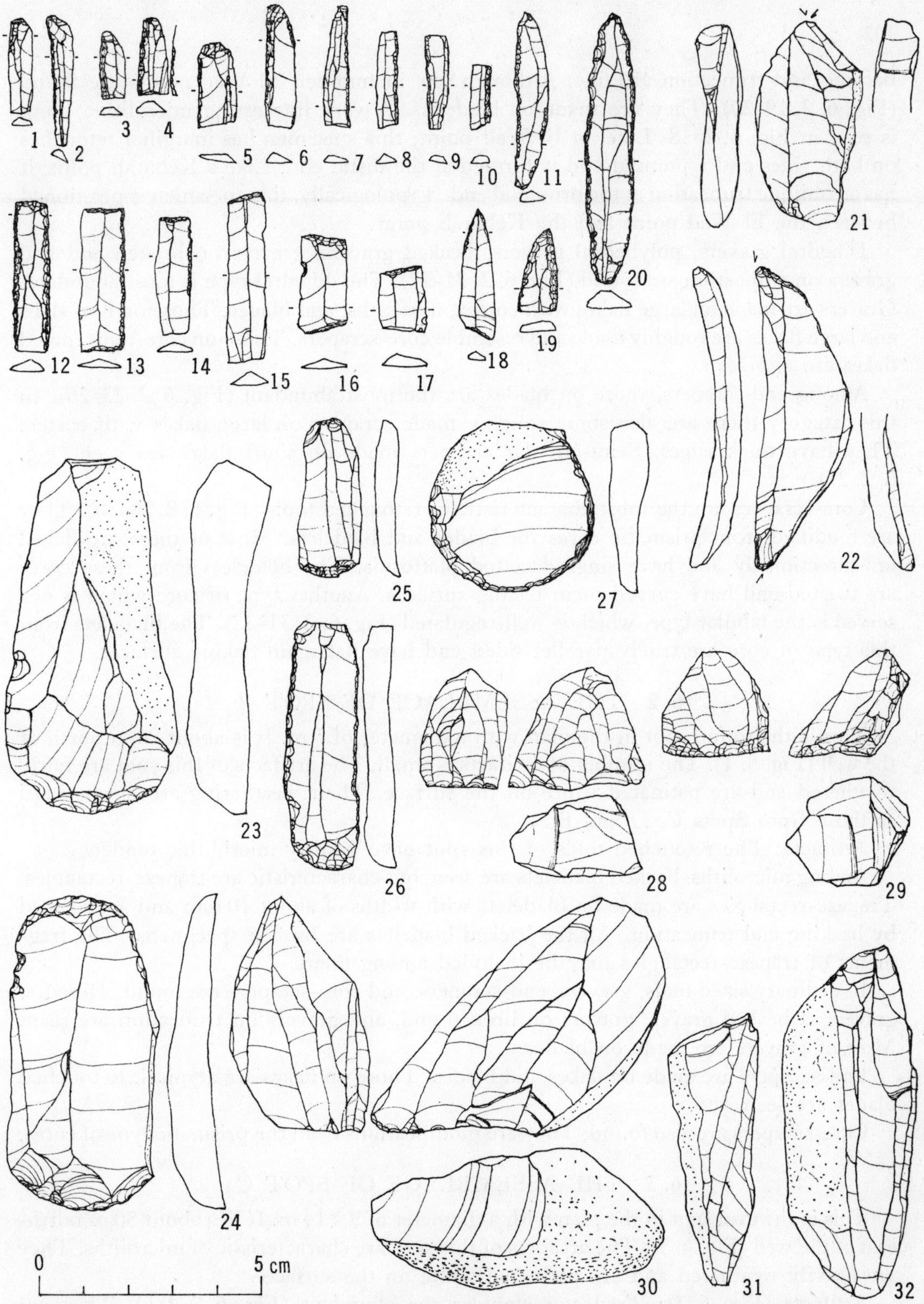

Fig. 6. 2. The stone tools of Spot A. 1–11: backed bladelets, 12–17: truncated bladelets, 18: microlithic point, 19–20: El Wad points, 21–22: gravers, 23–26: end-scrapers, 27: semi-circular scraper, 28–30: core-scrapers, 31–32: tabular cores.

backing and truncation are fine. Although few in number, El Wad points are found (Fig. 6. 2. 19–20). They are made on bladelets. A very interesting microlithic point is seen in Fig. 6. 2. 18. Like an El Wad point, this specimen has marginal retouches on both sides and a pointed end is formed at the distal end. Like a Kebarah point, it has an oblique truncation at the proximal end. Typologically, this specimen is positioned between the El Wad point and the Kebarah point.

Dihedral gravers, polyhedral gravers, beaked gravers, gravers on broken end and gravers on truncation are found (Fig. 6. 2. 21–22). The dihedral type is most abundant. Gravers are made on large flakes with cortex, flint slabs, and blades. Those on flint slabs and large flakes are roughly made and resemble core-scrapers. Those on core-rejuvenated flakes are also seen.

Among end-scrapers, those on blades are the most abundant (Fig. 6. 2. 23–26). In this category there are also some roughly made scrapers on large flakes with cortex. They have thick edges. Semi-circular scrapers made on short flakes are seen (Fig. 6. 2. 27).

Core-scrapers are the most characteristic of retouched tools (Fig. 6. 2. 28–30). They are modified from prismatic cores for blades and bladelets. Most of them are flaked uni-directionally and have single-facetted platforms. The bladelets from these cores are twisted and have curved main flaking surfaces. Another type of core which is observed is the tabular type, which is well-regulated (Fig. 6. 2. 31–32). The bladelets from this type of core are truely parallel sided and have flat main flaking surfaces.

6. 2 THE ASSEMBLAGE OF SPOT B

This is the round spot in the plan with a diameter of 5 m. It is about 20 m north of the well (Fig. 6. 1). The number of artifacts is small. The artifacts of this spot are fairly weathered and are patinated white on the surface. Their weathering are not so hard as those from Spots C, D and E.

Artifacts: The retouched tools of this spot have a fairly microlithic tendency.

Among microliths, backed bladelets are seen, but characteristic are trapeze-rectangles. Trapeze-rectangles are made on bladelets with widths of about 10 mm and are formed by backing and truncation. All the backed bladelets are broken specimens. The fragments of trapeze-rectangles may be included among them.

In ordinary sized tools, gravers, end-scrapers, and core-scrapers are found. Dihedral gravers, a beaked graver, gravers on broken end, and gravers on truncation are seen. Most of gravers are made on blades.

End-scrapers are made on flakes and blades. Those on flakes are atypical. Retouched blades are seen, too.

Core-scrapers are also found. They are modifications from the prismatic type of cores.

6. 3 THE ASSEMBLAGE OF SPOT C

This is the oval spot in the plan with a diameter of 9×14 m. It lies about 50 m northeast of the well (Fig. 6. 1). The artifacts of this spot are characteristic of microliths. They are heavily weathered and are patinated white on the surface.

Artifacts (Fig. 6. 3): Trapeze-rectangles are abundant (Fig. 6. 3. 1–6). They are definitely different from the truncated bladelets from Spot A, both technologically and

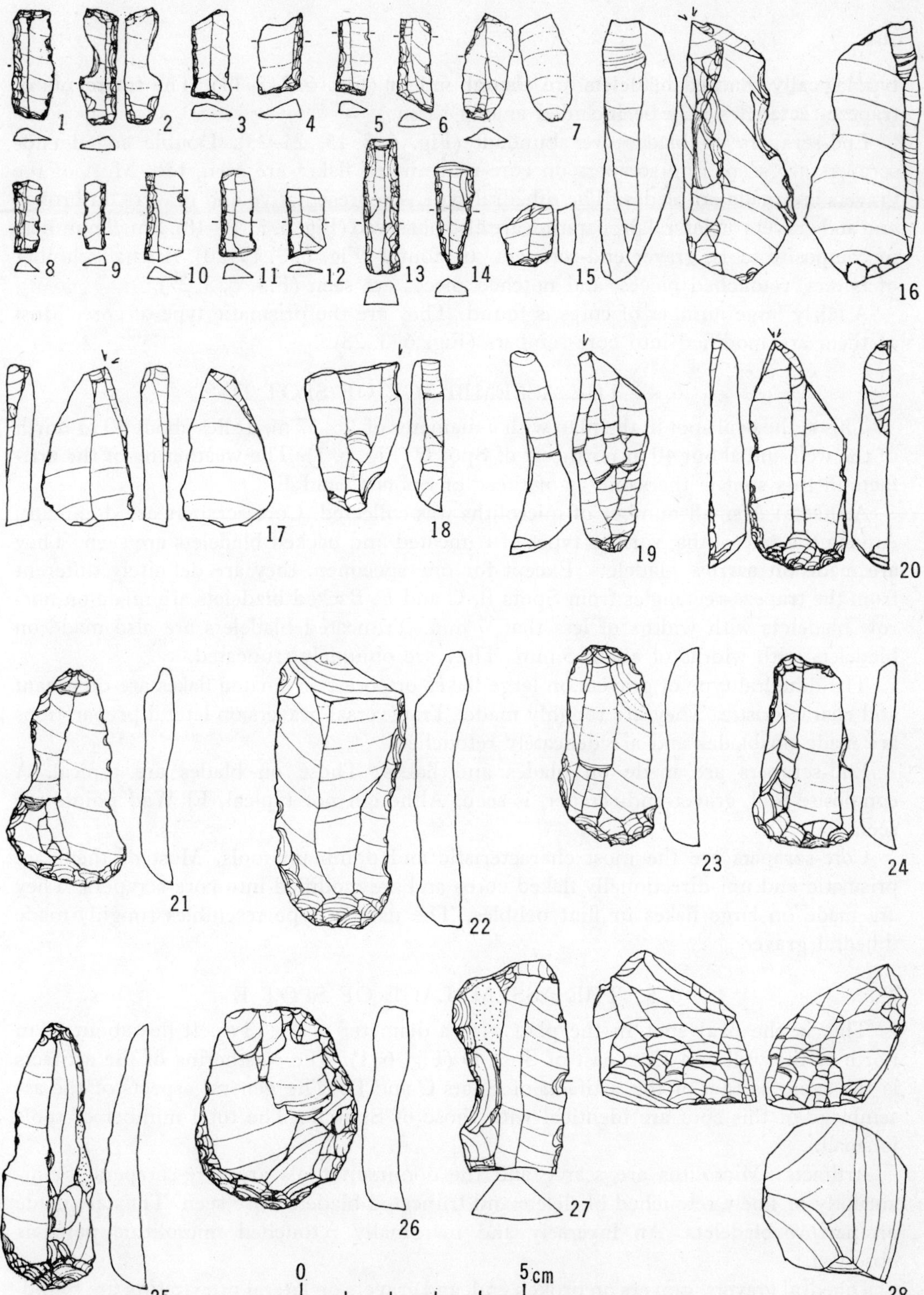

Fig. 6. 3. The stone tools of Spot C. 1–6: trapeze-rectangles, 7: micro-burin?, 8–14: backed bladelets, 15: end-scraper on bladelet, 16–18: gravers, 19–20: composite tools, 21–25: end-scrapers on blade, 26: semi-circular scraper, 27: notched blade, 28: core-scraper.

typologically. Backed bladelets are also abundant (Fig. 6. 3. 8–14). The fragments of trapeze-rectangles may be included among them.

End-scrapers on blades are abundant (Fig. 6. 3. 15, 21–25). Double edged end-scrapers are seen. End-scrapers on core-rejuvenated flakes are seen, too. Most of the gravers are made on blades. The dihedral type is representative and gravers on broken end and gravers on lateral preparation are also observed (Fig. 6. 3. 16–18). Small numbers of composite tools, graver-end-scrapers, are found (Fig. 6. 3. 19–20). A small number of borers, retouched pieces, and notched pieces are seen (Fig. 6. 3. 27).

A fairly large number of cores is found. They are the prismatic type of core. Most of them are modified into core-scrapers (Fig. 6. 3. 28).

6. 4 THE ASSEMBLAGE OF SPOT D

This is the oval spot in the plan with a diameter of 9×17 m. It lies about 60 m north of the well and about 40 m northeast of Spot C (Fig. 6. 1). The weathering of the artifacts in this spot is the same as of those of Spots C and E.

Artifacts: A small number of microliths was collected. Core-scrapers are dominant. Among the microliths, various types of truncated and backed bladelets are seen. They are made on narrow bladelets. Except for one specimen, they are definitely different from the trapeze-rectangles from Spots B, C and F. Backed bladelets are made on narrow bladelets with widths of less than 7 mm. Truncated bladelets are also made on bladelets with widths of about 5 mm. They are obliquely truncated.

The dihedral type of gravers on large flakes or core-rejuvenated flakes are dominant and characteristic. They are roughly made. Transversal gravers on lateral preparations are made on blades and are delicately retouched.

End-scrapers are made on blades and flakes. Those on blades are typical. A composite tool, graver-end-scraper, is seen. Although not typical, El Wad points are seen.

Core-scrapers are the most characteristic and dominant tools. Most of them are prismatic and uni-directionally flaked cores and are modified into core-scrapers. They are made on large flakes or flint pebbles. The narrow type resembles roughly-made dihedral gravers.

6. 5 THE ASSEMBLAGE OF SPOT E

This is the oval spot in the plan with a diameter of 7×13 m. It lies about 70 m north of the well and 15 m east of Spot D (Fig. 6. 1). The weathering of the artifacts is the same as that of the artifacts of Spots C and D. The general aspects of the assemblage of this spot are identical with those of Spot D. The total number of tools is small.

Artifacts: Microliths are scarce and the dominant tools are core-scrapers. Semi-abruptly or finely retouched bladelets and truncated bladelets are seen. They are made on narrow bladelets. An inversely and marginally retouched microlithic point is observed.

Dihedral gravers, gravers on broken end, and gravers on lateral preparation are found. The dihedral type is dominant. Among end-scrapers, those on blades are common. A borer and retouched pieces are also seen.

Core-scrapers are the most characteristic and dominant tools, as at Spot D. They are made on large flakes or flint pebbles.

6.6 THE ASSEMBLAGE OF SPOT F

The artifacts were concentrated in a small restricted area. At this spot, mapping and careful collection were not carried out. It lies about 200 m east of the well and on the edge of the terrace. The weathering of the artifacts is the same as that of the artifacts of Spot B. Half of the tools are microlithic.

Artifacts: Among microliths, trapeze-rectangles are characteristic. All of them are typical trapeze-rectangles. Backed bladelets are made on fairly broad bladelets.

Gravers are on broken end and on lateral preparation types. End-scrapers on blades are common. Otherwise, prismatic and tabular types of cores are seen.

6.7 CLASSIFICATION

The assemblages of Spots B, C and F are characteristic with a high proportion of microliths. Among microliths, trapeze-rectangles are abundant. The assemblages of Spots B, C and F are assigned to Geometric Kebaran A of Bar-Yosef (Bar-Yosef, 1970, 1975, 1976). According to Bar-Yosef (1976, pp. 95–97), the later stage of Geometric Kebaran A has a wider type of trapeze-rectangle with an abrupt retouch. All of the trapeze-rectangles from Spots B, C and F have an abrupt retouch and are 6–10 mm in width. Therefore, the assemblages of Spots B, C and F can be assigned to the later stage of Geometric Kebaran A. The assemblages of Stratigraphical Units A and B of Horizon II of Douara Cave are also assigned to the later stage of Geometric Kebaran A. Slight differences are observed between these assemblages.

The assemblages of Spots A, D and E are characterized by an abundance of core-scrapers. The dominance of core-scrapers is the cultural index of stage V of the Upper Paleolithic in the Levant (Neuville, 1934, 1951), Atlitian (Garrod and Bate, 1937, p. 41), or Levantine Aurignacian Phase C (Copeland, 1975, p. 343). Various types of microliths are observed in the sites dated to stage V of the Upper Paleolithic in the Levant, such as layer 2 of Rock-shelter II in Jabrud (Rust, 1950, pp. 89–93), layer 9 or E of El Khiam (Echegaray, 1964, pp. 97–107; Perrot, 1951, pp. 139–147), and layer C of El Wad (Garrod and Bate, 1937, pp. 41–44). Garrod excluded microliths from the assemblage of layer C of El Wad as an intrusion from the overlying layer B (Garrod and Bate, 1937, p. 43). As Perrot pointed out (1951, p. 147), these microliths might not be intrusive specimens.

Among the tools of Spot A, actually perpendicular truncated bladelets are characteristic. In layer 8 of Rock-shelter III in Jabrud, Skiftian of Rust (1950, pp. 103–107), perpendicularly truncated bladelets were recognized. In this layer of Jabrud, El Wad points, core-scrapers and microliths were found. The assemblage of Spot A resembles the assemblage of layer 8 of Rock-shelter III in Jabrud. The microlithic point from Spot A has the character of both types of the El Wad point and the Kebarah point and can be regarded as a shifting type from the El Wad point to the Kebarah point. The El Wad point is the cultural index of stage III of the Upper Paleolithic in the Levant, and it is also observed in the assemblages of stage V. The Kebarah point is found in the assemblages of Non-Geometric Kebaran (stage VI). As Hours stated (Hours,

Copeland et Aurenche, 1973, p. 273), stage V of the Upper Paleolithic in the Levant or Atlitian is not yet well studied and the precise chronological position of Skiftian is not certain. Apart from whether Skiftian belongs to stage V or not, the stratigraphical relationships of Skiftian have been certified at Rock-shelter III in Jabrud, between Jüngaurignacien (stage V?) and Nebekian (Kebaran or stage VI). Thus, the assemblage of Spot A can be assigned to Skiftian of Rust or shifting stage between stages V and VI. Assemblages like Skiftian have not yet been found at other sites. It is probable that Skiftian is a merely regional facies around the Syrian Desert.

The assemblages of Spots D and E can be dated to stage V of the Upper Paleolithic in the Levant, because of the dominance of core-scrapers and dihedral gravers.

The analysis of the assemblages is still going on and this article is merely a brief preliminary report, so the conclusions are tentative.

LITERATURE CITED

Bar-Yosef, O. (1970) Prehistoric Sites near Ashdad, Israel. Palestine Exploration Quarterly, 1970 Jan.–June: 52–63.

Bar-Yosef, O. (1975) The Epipaleolithic in Palestine and Sinai. In: Wendorf and Marks **1975**: 363–378.

Bar-Yosef, O. (1976) A Note on the Geometric Kebaran A. Pre-conference Papers for IXth International Congress of Pre- and Proto-historic Science, Nice, Colloque **III**: 78–105.

Copeland, L. (1975) The Middle and Upper Paleolithic of Lebanon and Syria in the light of recent Research. In: Wendorf and Marks **1975**: 317–350.

Echegaray, J. G. (1964) Excavacions en la terraza de El Khiam, Jordania. I. Studio del yacimento y los nievels Paleolitico. Madrid.

Garrod, D. A. E. and Bate, D. M. A. (1937) The Stone Age of Mt. Carmel. I. Oxford.

Hours, F., Copeland, L. et Aurenche, O. (1973) Les industries paléolithiques du Proche-Orient, Essai de corrélation (1). L'Anthropologie, Tome **77**: 229–280.

Neuville, R. (1934) La préhistorique de Palestine. Revue Biblique, Tome **43**: 237–259.

Neuville, R. (1951) Le paléolithique et le mésolithique du désert de Judée. Archives de L'Institut de Paléontologie Humaine, Mémoire 24. Paris.

Perrot, J. (1951) La terrasse d'El-Khiam. dans Neuville **1951**: 134–178.

Rust, A. (1950) Die Höhlefunde von Jabrud (Syrien). Neumüster.

Wendorf, F. and Marks, A. E. (eds.) (1975) Problems in Prehistory: North Africa and the Levant. Dallas.

CHAPTER 7

Concluding Remarks

Yutaka Sakaguchi

Department of Geography, Faculty of Science, The University of Tokyo

We have obtained some interesting results concerning the paleogeography of the Palmyra region related to the Paleolithic site of the Douara Cave. These results are very important for Quaternary studies in the Syrian Desert. The author will try to pick up and discuss some common problems among the articles in our report and add some supplementary data and comments.

7. 1 ON THE CORRELATION OF THE TERRACES AND THEIR AGES

The author divided the lacustrine terraces around Sabkhet Mouh into 397 m-, 387 m-, and 383 m-terraces, based on a standard shoreline elevation. However, it is also possible to consider that the 397 m-terrace was lowered by wind erosion, thus changing into the 387 m-terrace.

The 397 m-terrace can be definitely correlated to the middle terrace (S_3) in the Douara Basin, because the deposits of both terraces include abraded Middle Paleolithic artifacts. The 397 m- and 383 m-terraces may be correlated to the middle terraces I and II respectively in the Douara Basin.

Akazawa found Late Acheulean artifacts on the surface of the terrace of Locality 72 in the Palmyra Basin, which is higher than the 397 m-terrace. This terrace is correlated to the higher terrace (S_2) in the Douara Basin. The terrace which is located south of Locality 72 and higher than the terrace of Locality 72 may be correlated to the highest terrace (S_1) in the Douara Basin.

According to Koizumi, the general topography of the Douara Basin was formed in the time of the higher terrace, that is, before the Middle Paleolithic age. This conclusion may be also applied to the case of the Palmyra Basin. The lower terrace in the Palmyra Basin was inferred to have been formed during the minor high level stage of the pluvial lake, which was correlated to the Epi-Paleolithic and Pre-Pottery Neolithic age. Therefore, this terrace can be definitely correlated to the lower terrace on which Pre-Pottery Neolithic artifacts were found.

Considering these results and the author's investigations at Wadi el Ahmar, Wadi Jaria, and Arak, it is concluded that the geomorphological succession is common in the Palmyra region.

7. 2 ON THE GENESIS OF THE TERRACE

The middle terrace in the Douara Basin is an accumulation terrace. The accumulation is inferred to have been brought about by two causes: an increase in transporting loads and a rise in the local base level of erosion, which was caused by the formation of the pluvial lake. If this inference is correct, the higher terrace in the Douara Basin and the correlated ones may offer a key to the solution of the problem of whether or not a lake existed in the time of the higher terrace. Therefore, the question of whether these terraces are filltop terraces or not is important. Koizumi considered that the higher terrace in the Douara Basin was an accumulation terrace. However, according to the author's investigations, the higher terrace in the Douara Basin is considered to be mostly a rock-cut terrace.

Koizumi considered that the down-cutting of the middle terrace surface was mainly caused by a decrease in transporting loads. However, it seems to be reasonable to conclude that the main cause was the lowering of the water level of the pluvial lake. The accumulation and the down-cutting of the lower terrace may also be explained by the minor fluctuations in the water level in the Epi-Paleolithic and the Pre-Pottery Neolithic ages.

7. 3 ON NEOTECTONICS DURING THE LATE QUATERNARY IN THIS REGION

There is no certain evidence concerning crustal movements related to the development of the landforms in this region. However, the following data are suggestive in connection with this problem.

Three terraces are developed along Wadi Jaria at the foot of the southern slope of the eastern Jabal ad Douara (Locality 43). The longitudinal profiles of the upper and the middle terrace strikingly diverge toward the mountain. However, the middle terrace is excluded from consideration, because this terrace is a filltop one, and it is difficult to decide whether or not the inclination of its profile is an original dip. The upper terrace is a rock-cut one. The inclination of its profile is too steep. This situation may be caused by the uplift of the mountain.

There are four terraces along Wadi el-Ahmar at Locality 44, 3 km southwest of the Douara Cave. The surface of the highest terrace slightly inclines toward the north. The direction of the inclination coincides with that of its bedrock, the Pliocene conglomerates. This fact may also suggest that crustal movements occurred in this area.

Accordingly, it is possible to infer that crustal movements had taken place at least before the Middle Paleolithic age in Jabal ad Douara area.

7. 4 ON THE ASYMMETRIC ARRANGEMENT OF THE TOPOGRAPHY IN THE DOUARA BASIN

As shown in Fig. 3. 5 in Koizumi's article, some landform elements arrange asymmetrically in the NS profile of the Douara Basin. The lowest part of the basin, which is through Wadi Khawabi el Kharrar and Wadi Jaria, is located near the southern margin of the basin. The middle terrace is mostly distributed north of this lowest part. The taluses are developed along the foot of the south-facing scarp surrounding the basin, while the distribution of the pediments are restricted to the foot of the north- and west-facing scarp. As Koizumi pointed out, this asymmetry seen in the pediment-talus system is considered to have been caused by differential debris production depending on the direction of the scarp.

Though Koizumi inferred that the debris production was controlled by the dip of the strata consisting of the scarp surrounding the basin, the author considers that the asymmetry in this basin can be explained by differential insolation between opposite scarps. Insolation is greater on the south-facing scarp, where weathering has been more intense. In contrast, on the north-facing scarp insolation is much less, and weathering has proceeded much more slowly. The west-facing scarp is in the intermediate state between these two cases. If runoff takes place uniformly in a basin, produced debris is washed out from the whole area. Consequently, pediments are easily formed at the foot of the north-facing scarp, but debris remains to form taluses at the foot of the south-facing scarp. A fresh rock wall appears and then becomes more disintegrated. This process is repeated, and the south-facing scarp retreats rapidly. Why are taluses poor at the foot of the southern slope of Jabal ad Douara? Because this slope is a dip slope, and not so much debris is produced.

7. 5 ON THE PROBLEM OF LOCALITY 50

As previously mentioned, there is a discrepancy between the stratigraphy of the deposits seen on the well and the archeological spots found on the ground around the well. According to Fujimoto, the ages of these spots decrease in the following order: D and E (the stage V of the Upper Paleolithic age in Levant)—A (the well, the transitional stage V / VI of the Upper Paleolithic age in Levant)—B, C, and F (the stage of the Geometric Kebaran A). Spots D and E are located on the gentle slope between 385 m and 386 m in elevation. Spots B, C, and F are located on the nearly horizontal ground between 384 m and 385 m in elevation. The artifacts-bearing strata at Spot A are considered to be secondary in origin. If the following inference is correct, the relation between the ages of the spots and the topography may be explained without contradiction. The artifacts of Spot A were transported from a primary spot located on the slope where Spots D and E are situated, then deposited at Spot A. Later, Spots B and C were located on the ground immediately north of Spot A.

The crust around Palmyra does not contain any artifacts more recent than the Epi-Paleolithic. However, the ground of Spots B to F is not encrusted. The result obtained from the Ghab pollen diagram and the deposition of the Fan gravels I and II at the foot of Douara Cave obtained by K. Endo's investigation are suitable for considering that the Pre-Pottery Neolithic age was humid and the crust around Palmyra was formed during this time. However, this consideration is contradictory to the situation of Locality 50. The crust formation may have taken place twice or more times.

7. 6 ON THE PALEOCLIMATE RELATED TO THE PALEOLITHIC SITE OF DOUARA CAVE

Among eleven radiocarbon dates of the samples obtained from Douara Cave in the 1974 season, ten dates are over the determination limit of 43,000 y. (TK) or 52,000 y. (GrN). Meanwhile, the travertine taken from the ground of the innermost minor cave of Douara Cave was dated at 70,000 F.T.y. by M. Suzuki (personal communication). Considering these dates, the age of the cave is surely more than 50,000 years.

According to Suzuki, the gypsum crystal taken from the deposits of the 397 m-terrace was dated at 70,000 y.B.P. by the fission-track dating method. This date supports the author's inference that the Paleo-Palmyra Lake is older than 45,000 years.

From the investigations independently carried out by each investigator in the Palmyra and Douara Basins, we have reached the same conclusion on the paleoclimate related to the Paleolithic site of Douara Cave. Summarizing our results, the Middle Paleolithic age in this region was rainy and cool in the summer, and in the winter colder than today. In the last glacial period, the snowline in the Lebanese Mountains is estimated to have been lowered by 1050 to 1350 (?) m (Kaiser, 1963) or 1000 m (Messerli, 1967). If the lowering of the snowline depended on the temperature only, the lowering of the temperature in the last glacial period may be 5.5 to 7.5°C. Recently the lowering of the temperature by 15°C has been inferred by an investigation at Mt. Hermon, the Anti-Lebanese Mountains (Gvirtzman, 1976).

LITERATURE CITED

Gvirtzman, G. (1976) Late Würm temperature depression in the Middle East −15°C: Evidence from fossil snowlines on Mount Hermon and Jebel Catharina, Sinai. Proceedings IGC, Moskva, **2**: 295–297.

Kaiser, K. (1963) Die Ausdehnung der Vergletscherung und "periglazialen" Erscheinungen während der Kaltzeiten des quartären Eiszeitalters innerhalb der syrisch-libanesischen Gebirge und die Lage der klimatischen Schneegrenze zur Würmeiszeit im östlichen Mittelmeergebiet. Proceedings INQUA, Warsaw, 127–148.

Messerli, B. (1967) Die eiszeitliche und die gegenwärtige Vergletscherung im Mittelmeerraum. Geographica Helvetica, **22**: 105–228.

APPENDIX

Petrological Study of Flints from the Douara Basin, Northeast of Palmyra

Yasuji Saito and Tokiko Tiba
Department of Geology, National Science Museum

It is well known that flints or nodular cherts occur in sedimentary carbonate rocks. Most of such siliceous nodules originate from replacement of carbonate minerals by silica (Emery and Rittenberg, 1952; Rutten, 1957; Pittman, 1959; Kanmera, 1968; etc.). The origin of silica, however, has long been a controversial subject. Debate has centered on whether the silica is biogenic or of inorganic precipitation. Flints from the Douara Basin, northeast of Palmyra, Syria, throw some light on the problem, as the flints display many apparently primary sedimentary textures.

The marl and limestone of the Douara Basin are dense and commonly foraminiferous. They are sometimes silicified and phosphotized, and contain abundant flints and barite nodules in some horizons of the thick sequence (Soulidi-Kondratiev, 1966). The flints, forming nodules or relatively thin beaded bands, were used as stone wares. In this paper, the flints are described and the possible carbonate-silica replacement origin of flints is discussed.

STUDY METHODS

The study is divided into three parts: a conventional petrographic examination of thin sections to observe the constituent minerals and the textures of the flints; X-ray diffraction and chemical analyses for bulk rock mineralogy and chemistry; and scanning electorn microscope observations. The last attempt was made to investigate the primary features expected to be preserved in the flints rather than the secondary mineralogical features or textures by recrystallization, as scanning electron microscopy is very useful in observing morphological features of constituent materials of other siliceous sedimentary rocks (Imoto and Saito, 1973; Tiba, Matsubara and Saito, 1976).

MATERIAL

The flints studied are mostly lens-formed nodules approximately 20 cm in diameter and 10 cm in thickness. They have a dense and subvitreous appearance but vary in color. The commonest color is light brown to brown, and, in relatively larger blocks of the flints, dark brown or black. The rough crust, a few centimeters thick, is white or light-colored. To the naked eye, some flints may often be mistaken for bedded chert or obsidian because of their subvitreous luster and subconcoidal fracture.

PBA-1 (Fig. 1. a): A fragment of nodule with a 2 mm thick crust and light brown gritty surface. The inner part shows a milky cocoa color, vitreous luster, and uneven angular fracture relief.

PBA-2 (Fig. 1. b): A lens-formed nodule with a light gray crust of 1–1.5 cm thickness. The surface of the crust is gritty to the touch. The inner part is light cocoa in color and of unglazed porcelain luster. The part closer to the crust is rather dark.

PB-35 (Fig. 1. c): A nodule with a 1–3 mm thick crust. The surface of the crust is subgritty. The boundary between the crust and the inner part is often indistinct. The inner part is further subdivided into the outer raw amber portion and the inner black portion. The boundary between the two is irregular and gradual. Both portions show a subvitreous luster and concoidal fracture similar to obsidians.

THIN SECTION STUDY

In thin sections, the flints are light to medium brown translucent chert consisting of fine-grained equidimensional quartz and chalcedonic quartz with minor amounts of carbonate (?) minerals. Abundant foraminiferal pseudomorphs are scattered (Photos 1–1a, 1–2a, 2–1a and 2–2a). Although the foraminiferal tests are replaced and their cavities are filled with silica, the forms of those microfossils seem to be well preserved in the flints (Fig. 2). It is suggested that the foraminiferal tests did not dissolve until silica precipitated in their cavities and in cementing groundmass and that the replacement of carbonate by silica progressed gradually. The foraminiferal shells consist of cryptocrystalline to microcrystalline quartz, which is indistinguishable from the surrounding cement material. The material filling the foraminiferal pore spaces is mostly coarser-grained, colorless chalcedonic quartz (Photos 1–1b, 1–2b, 2–1b, and 2–2b). Siliceous sponge spicules are also common in the flints as aggregates of colorless cryptocrystalline quartz but such fine textures as central canals of spicules are not preserved.

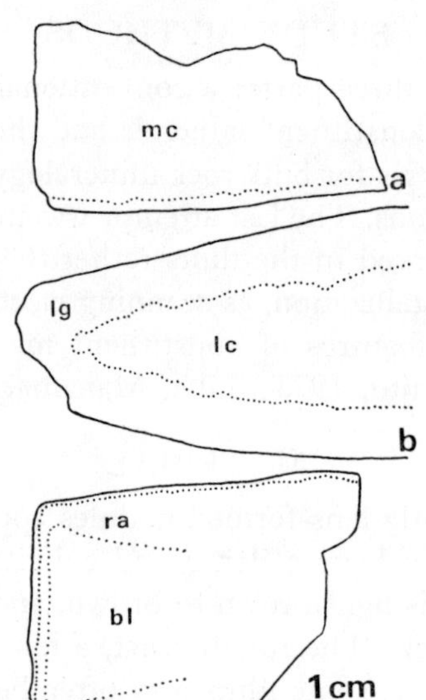

Fig. 1. Sections of the examined flints.
a: PBA-1, b: PBA-2, c: PB-35. mc: milky cocoa color, lg: light gray, lc: light cocoa, ra: raw amber, bl: black.

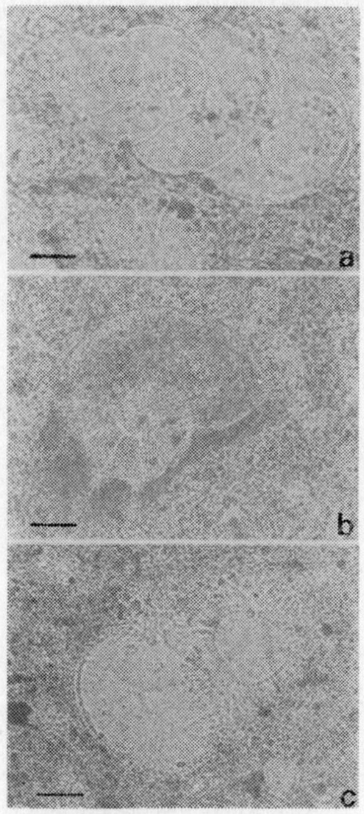

Fig. 2. Microphotographs of well-preserved foraminiferal tests in the flints. Open nicol, length of bars 0.05 mm. a and b: benthonic, c: planktonic(?).

Some flints contain rhombohedral crystals of carbonate in appreciable amounts. Opaque minerals such as pyrite and its weathering products are common minor constituents of the flints.

PBA-1 (Photos 2–2a and 2–2b): Foraminiferal shells, 0.05 to 0.3 mm in diameter, constitute less than 50% in volume of the bulk rock. The foraminiferal cavities consist mainly of radial aggregates of fibrous chalcedonic quartz, whose grains are coarser than those of the enclosing microcrystalline quartz in the matrix. The tests of foraminifers are replaced by fine-grained fibrous chalcedonic quartz growing approximately perpendicularly to the inner or outer walls of the tests. Refractive indices of those quartz crystals are higher in the test than in the cavities. Chalcedonic fibers are length-fast. The microcrystalline matrix consisting of more than 50% in volume of the bulk rock contains abundant spicules less than 0.2 mm in length. Network veins of quartz often cut the foraminifers. Opaque mineral (possibly pyrite) is also present.

PBA-2 (Photos 2–1a and 2–1b): This specimen contains smaller amounts of foraminifers than the specimen PBA-1. The foraminiferal cavities are filled with chalcedonic quartz or with material of light-brown color in plain light but no double refraction. Other features are similar to those of the specimen PBA-1 excluding the absence of quartz veins.

PB-35 (Photo 1): The foraminiferal cavities are filled not only with clear chalcedonic quartz but also with brown translucent or opaque materials. In the enclosing matrix of microcrystalline quartz are apatite-like grains less than 0.15 mm long. Abundant spicules are contained. The other features are similar to PBA-1 and PBA-2.

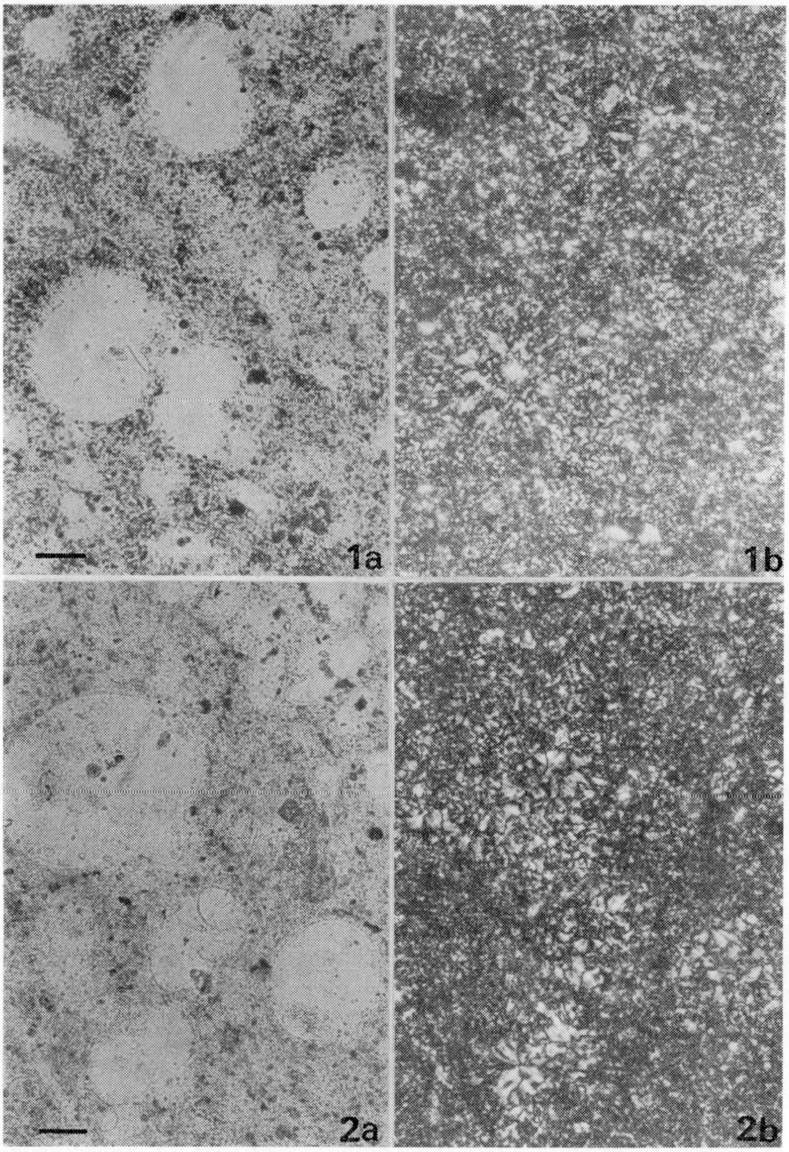

Photo 1. Photomicrographs showing replacement of foraminiferal tests by silica in the flint (PB-35) from the Douara Basin, Syria. This specimen contains numerous silicified foraminifers and their cavities are filled with relatively coarser-grained chalcedonic quartz than enclosing matrix of microcrystalline quartz.
a: open nicol, b: crossed nicols. Length of bars: 0.05 mm.

SCANNING ELECTRON MICROSCOPY

Before examination with the scanning electron microscope, all the specimens of the flints were etched by a solution of 10 to 20% hydrofluoric acid, then washed with running water and cleaned by means of ultrasonic vibration to remove dusty materials

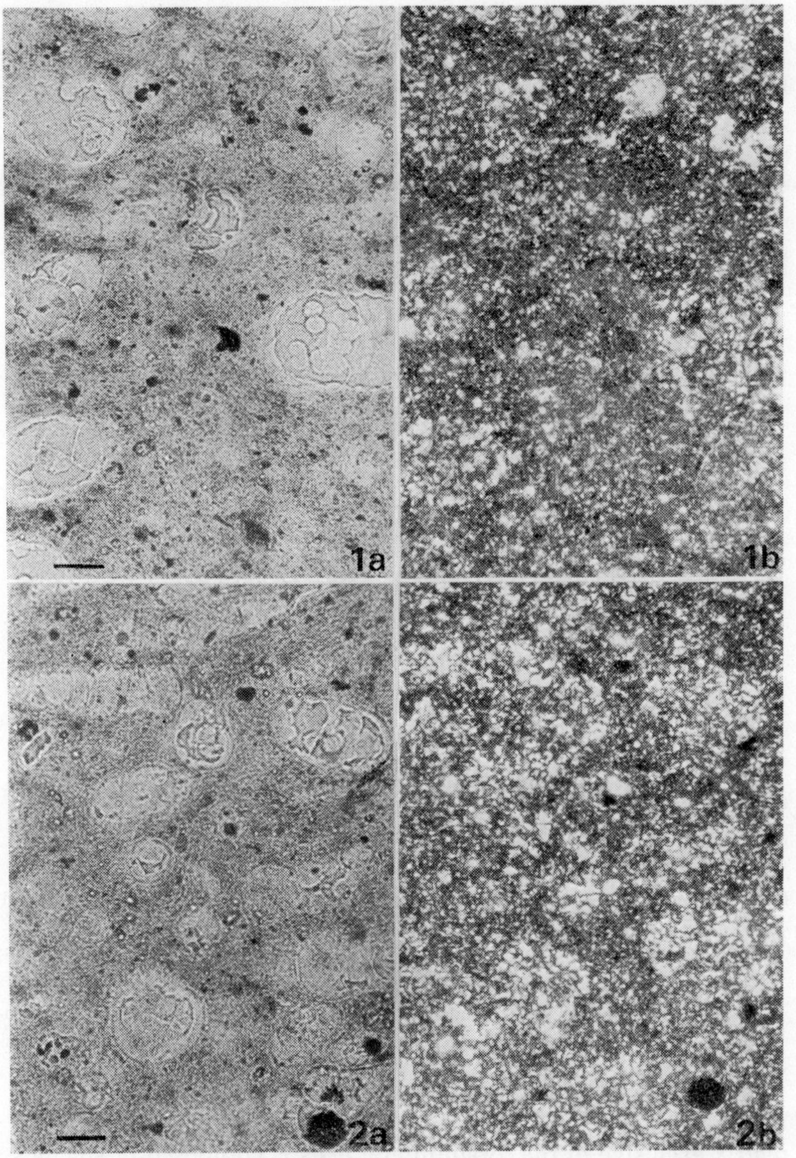

Photo 2. Photomicrographs showing replacement of foraminiferal tests by silica in the flints from the Douara Basin, Syria.
a: open nicol, b: crossed nicols, 1: PBA-2, 2: PBA-1.
Length of bars: 0.1 mm.

on the etched surfaces. The etched chips were coated with gold in a vacuum evaporator, and then examined.

Observation under the scanning electron microscope distinctly shows that the flints from the Douara Basin consist mainly of silicified foraminifers, spines or spicules, and their fragments (Photos 3–1 and 3–2). The spicules seem to be arranged with preferred orientation (Photo 3–2). Such texture probably reflects a condition at the time of deposi-

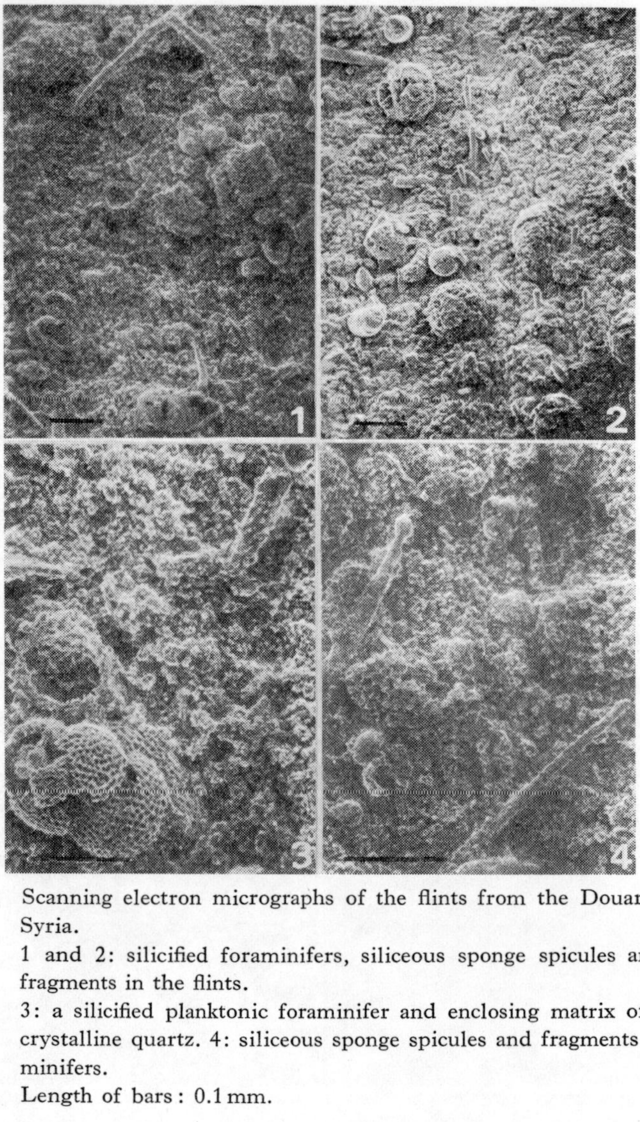

Photo 3. Scanning electron micrographs of the flints from the Douara Basin, Syria.
1 and 2: silicified foraminifers, siliceous sponge spicules and their fragments in the flints.
3: a silicified planktonic foraminifer and enclosing matrix of microcrystalline quartz. 4: siliceous sponge spicules and fragments of foraminifers.
Length of bars: 0.1 mm.

tion. The matrix exclusive of the foraminiferal tests is composed of spicules, fragments of the tests, and spongy microcrystalline quartz (Photos 3–3 and 3–4). The magnified spongy parts show that the finer matrix is also of biogenic origin. Many microfossils, such as fine spicules with overgrowth by recrystallization (Photos 4–1 and 4–2), coccoliths replaced by silica (Photos 4–1 and 4–3), and an unidentified one (Photo 4–4), can be obviously recognized. These facts indicate that the constituent material of the flints deposited is an ill-sorted mixture of calcareous and siliceous microfossils. The flints are, therefore, not the products of inorganic precipitation of silica during deposition, but are the products of epigenetic replacement by silica.

Chalcedonic quartz filling a foraminiferal cavity and a part of a foraminiferal test replaced by cryptocrystalline quartz are shown in Fig. 3. The chalcedonic quartz seems

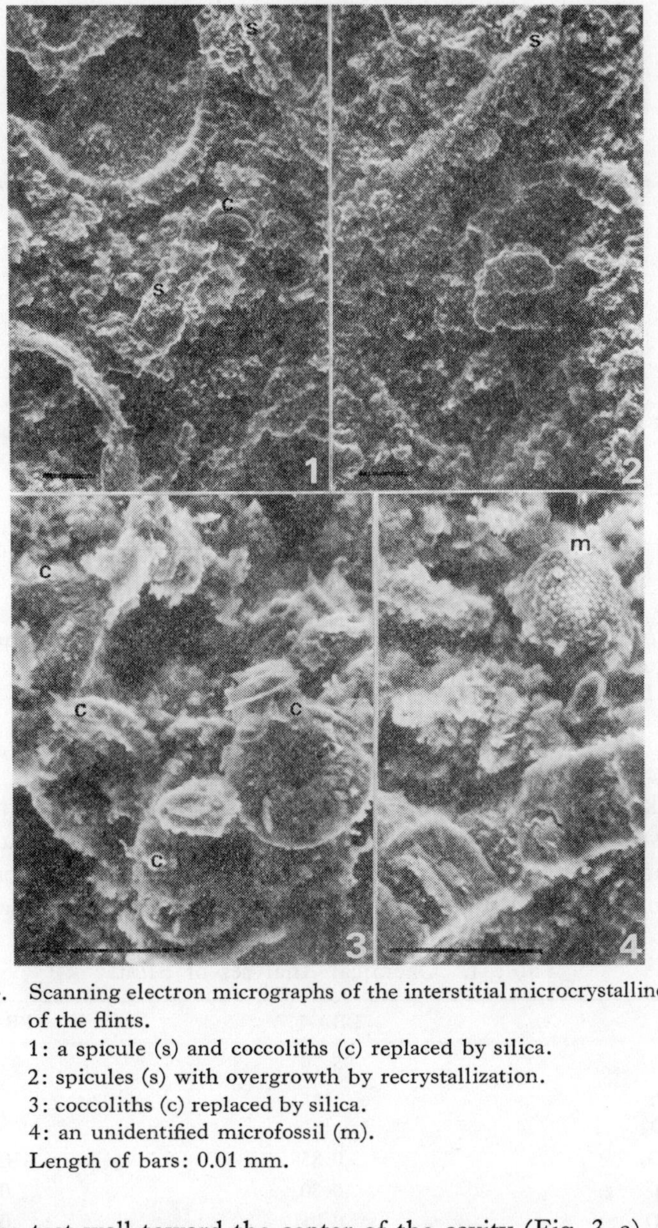

Photo 4. Scanning electron micrographs of the interstitial microcrystalline quartz of the flints.
1: a spicule (s) and coccoliths (c) replaced by silica.
2: spicules (s) with overgrowth by recrystallization.
3: coccoliths (c) replaced by silica.
4: an unidentified microfossil (m).
Length of bars: 0.01 mm.

to grow from the test wall toward the center of the cavity (Fig. 3. a). This may mean that the foraminiferal pore space had not yet been filled with fine-grained clastic material at the time of deposition. In the foraminiferal test, most of the very fine-grained quartz crystals are xenomorphic (Fig. 3. b). The irregular forms and the fine size of the crystals may support relatively slow and gradual dissolution of calcite tests and concomitant precipitation of silica. Rapid dissoltion of the tests would have formed relatively wide pore space, where larger crystals may be able to precipitate.

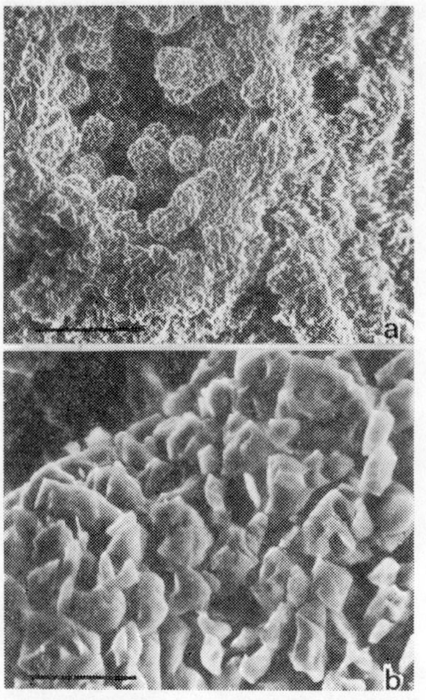

Fig. 3. Chalcedonic quartz filling foraminiferal cavity (a) and a part of foraminiferal test replaced by cryptocrystalline quartz (b).
Length of bars : 0.02 mm (a), 0.002 mm (b).

X-RAY POWDER DIFFRACTION DATA

The bulk rock samples of the flints examined here consist almost entirely of quartz, in some cases with minor amounts of carbonate minerals. These results are not contradictory to those of the conventional petrographic study. The absence of distict patterns of carbonate minerals in X-ray diffraction analyses suggests that rhombohedral

Table 1. Chemical Analyses of Flints

	PBA-1	PBA-35
SiO_2	95.70	93.58
TiO_2	tr	tr
Al_2O_3	tr	tr
Fe_2O_3	0.85	0.42
FeO	0.30	0.37
MnO	0.01	0.01
MgO	tr	tr
CaO	1.08	1.49
Na_2O	0.08	0.16
K_2O	0.08	0.10
H_2O-	1.53	3.36
H_2O+	0.26	0.22
P_2O_5	0.36	0.57
Total	100.25	100.28

crystals observed in thin sections are pseudomorphs of original constituent minerals.

CHEMICAL ANALYSES

Two bulk rock samples were chemically analyzed (Table 1). In spite of differences in appearance (color, luster, fracture etc), these flints have quite similar compositions with only significant differences in SiO_2 and H_2O^+ contents. The flints with obsidian-like appearance in which foraminifers are ill-preserved (PB-35) are richer in SiO_2 and poorer in H_2O^+ than the chert-like ones (PBA-1), though very high SiO_2 content is common to both. Lack of Al_2O_3 and MgO means that these flints do not contain clay minerals in appreciable amounts. This is consistent with the absence of clay minerals confirmed by petrographic and X-ray studies and suggests that terrigenous materials had not been introduced at the time of formation of these flints. From these analyses, it is inferred that these flints are composed mainly of quartz with subordinate iron oxides and/or iron hydroxides and accessory apatite.

REPLACEMENT OF CARBONATE BY SILICA

Silica replacement of carbonate sedimentary rocks has been asserted on the basis of the presence of carbonate rhombs as remnants of host rocks, pseudomorphs of carbonate minerals within chert nodules, partial silicification of fossils across the boundary between carbonate and chert, and the persistence of textures such as laminations through chert nodules. The flints or nodular cherts from the Douara Basin, Syria, give direct evidence of a replacement origin. They consist of chalcedonic quartz and microcrystalline quartz. No calcite and dolomite are found and detrital material such as quartz, feldspar and clay minerals, is also absent. Most flints, however, contain abundant foraminiferal shells consisting of chalcedonic quartz. Pseudomorphs of coccoliths are also confirmed in the enclosing matrix of microcrystalline quartz. Such silicified foraminifers and coccoliths show that their original carbonate material was replaced by silica after deposition.

SILICA SOURCE

According to the geochemical studies, precipitation of silica from sea water is almost impossible by inorganic means because of its low concentration, so organic agents are the most likely source of silica precipitation (Correns, 1950; Degens and Epstein, 1962; Krauskopf, 1956, 1959; Siever, 1957). The presence of abundant siliceous spicules in the flints from the Douara Basin is a key to the solution of the problem of the silica source. Any silica supply of volcanic or hydrothermal and other inorganic origin cannot be expected in the Douara Basin.

In these flints, the original silica has been concentrated by organisms, in particular siliceous sponges, and has deposited, with abundant foraminifers, coccoliths and other carbonate materials (Fig. 4. a). The siliceous sponge spicules composed of low crystallized silica were buried and dissolved with increasing temperature and pH in the solution trapped in the sediments (Krauskopf, 1956, 1959; Okamoto et al., 1957; Siever, 1957). Provided that dissolution of sponge spicules is promoted, sponge spicules must be preserved only partly or consumed entirely. There are, however, abundant minute sponge spicules in these flints. It is probable that only the peripheries of the sponge

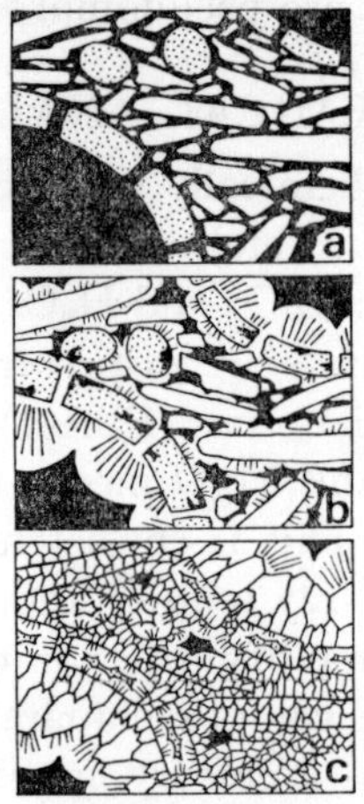

Fig. 4. Schematic formation process of the flints from the Douara Basin, Syria. a: unconsolidated sediment consisting of mixture of siliceous sponge spicules (white), foraminiferal tests and coccoliths (dotted). b: formation of chalcedony fibers and filling of free pore spaces (black). c: consolidation by dissolution of carbonate and concomitant precipitation of silica.

spicules were dissolved into solution and their main parts were left undissolved. This may be attributed to the pressure solution effect as in the case of quartzose sandstone (Thompson, 1959). Vast number of sponge spicules could have supplied considerable amount of silica sufficient to replace foraminifers and coccoliths. A decrease of pH has caused solution of the carbonate materials and precipitation of silica side by side. Carbonates were lost from the area under such conditions. Eventually the free pore spaces of foraminiferal cavities and the matrix were filled with silica (Fig. 4. b). Formation of chalcedony fibers coarser than those of microcrystalline matrix was possible only in the larger free space of foraminiferal cavities. The solution and precipitation of silica in pore spaces must have resulted in a reduction in porosity and bulk volume. The well-preserved foraminiferal pseudomorphs and their fine-grained xenomorphic quartz crystals probably imply a slow and gradual process of replacement. Abundant chalcedony overgrowth on spicules suggests that there were both pore spaces and the silica-saturated intersticial solution. The amorphous silica filling the pore spaces has been turned into chalcedonic and microcrystalline quartz through diagenesis (Fig. 4. c).

Acknowledgments

The authors are indebted to Mr. T. Akazawa, Department of Anthropology, National Science Museum, for providing flint samples, and to Drs. M. Hashimoto, A. Kato, Y. Kuwano, and S. Matsubara, Department of Geology of the same Museum, for their help in preparing the paper and X-ray diffraction analyses of flints.

LITERATURE CITED

Correns, C. W. (1950) Zur Geochemie der Diagenese. Geochim. et Cosmochim. Acta, **1** (**1**): 49–54.

Degens, E. T. and Epstein, S. (1962) Relationship between O^{18}/O^{16} Ratios in Coexisting Carbonates, Cherts, and Diatomites. Amer. Assoc. Petroleum Geologists Bull., **46** (**4**): 534–542.

Emery, K. O. and Rittenberg, S. C. (1952) Early Diagenesis of California Basin Sediments in Relation to Origin of Oil. Ibid., **36** (**5**): 735–806.

Imoto, N. and Saito, Y. (1973) Scanning Electron Microscopy of Chert. Bull. Natn. Sci. Mus., **16** (**2**): 397–400.

Kanmera, K. (1968) On Some Sedimentary Rocks Associated with Geosynclinal Volcanic Rocks. Mem. Geol. Soc. Japan, (**1**): 23–32 (in Japanese).

Krauskopf, K. B. (1956) Dissolution and Precipitation of Silica at Low Temperature. Geochim. et Cosmochim. Acta, **10** (**1**): 1–26.

Krauskopf, K. B. (1959) The Geochemistry of Silica in Sedimentary Environments. Soc. Econ. Paleontologists and Mineralogists, Spec. Pub., (**7**): 4–19.

Okamoto, G., Okura, T., and Goto, K. (1957) Properties of Silica in Water. Geochim. et Cosmochim. Acta, **12** (**1**): 123–132.

Pittman, J. S. (1959) Silica in Edwards Limestone, Travis County, Texas. Soc. Econ. Paleontologists and Mineralogists, Spec. Pub., (**7**): 121–134.

Rutten, M. G. (1957) Remarks on the Genesis of Flints. Amer. Jour. Sci., **255** (**6**): 432–439.

Siever, R. (1957) The Silica Budget in the Sedimentary Cycle. Amer. Mineralogist, **42** (**11 & 12**): 821–841.

Soulidi-Kondratiev, E. D. (Compiled) (1966) Explanatory Notes of the Geological Map of Syria. Sheet J-37-XV (Tudmor). Dept. Geol. Mineral Research, Syrian Arab Republic.

Thompson, A. (1959) Pressure Solution and Porosity. Soc. Econ. Paleontologists and Mineralogists, Spec. Pub., (**7**): 92–110.

Tiba, T., Matsubara, S., and Saito, Y. (1976) Nodular Diatomite from Dogo, Oki Islands. Bull. Natn. Sci. Mus., Ser. C (Geol.), **2** (**1**): 67–70.

